Communications in Computer and Information Science 1379

T0171858

More information about this series at http://www.springer.com/series/7899

Tristan Cazenave · Olivier Teytaud ·
Mark H. M. Winands (Eds.)

Monte Carlo Search

First Workshop, MCS 2020
Held in Conjunction with IJCAI 2020
Virtual Event, January 7, 2021
Proceedings

 Springer

Editors
Tristan Cazenave
Université Paris-Dauphine
Paris, France

Olivier Teytaud
Facebook FAIR
Paris, France

Mark H. M. Winands
Maastricht University
Maastricht, Limburg, The Netherlands

ISSN 1865-0929   ISSN 1865-0937 (electronic)
Communications in Computer and Information Science
ISBN 978-3-030-89452-8   ISBN 978-3-030-89453-5 (eBook)
https://doi.org/10.1007/978-3-030-89453-5

This Springer imprint is published by the registered company Springer Nature Switzerland AG
The registered company address is: Gewerbestrasse 11, 6330 Cham, Switzerland

Preface

These proceedings contain the papers presented at the first Monte Carlo Search Workshop, which was held virtually as an IJCAI workshop on January 7, 2021.

Monte Carlo search is a family of search algorithms that have many applications in different domains. It is the state of the art in many perfect- and imperfect-information games. Other applications include the RNA inverse folding problem, logistics, multiple sequence alignment, general game playing, puzzles, 3D packing with object orientation, cooperative pathfinding, software testing and heuristic model checking.

In recent years, many researchers have explored different variants of Monte Carlo search, their relationship to deep reinforcement learning, and their different applications. The purpose of the workshop was to bring these researchers together to present their research, discuss future research directions, and cross-fertilize the different communities. Submissions were welcome in all fields related to Monte Carlo search, including:

- Monte Carlo tree search and upper confidence trees,
- Nested Monte Carlo search,
- Non-locality in Monte Carlo search,
- Combination with zero learning,
- Monte Carlo belief-state estimation,
- Self-adaptive Monte Carlo search,
- Monte Carlo in many player games,
- Industrial applications,
- Scientific applications,
- Applications in games,
- Applications in puzzles.

Each paper was sent to three reviewers. If conflicting views on a paper were reported, it was sent to an additional reviewer. Program chairs were not involved in the review assignment and final selection of their own papers as these were handled separately by the other chairs. In the end, 11 contributions were accepted for presentation at the workshop, of which nine made it into these proceedings. Here we provide a brief outline of these nine contributions, in the order in which they appear in the proceedings.

"The $\alpha\mu$ Search Algorithm for the Game of Bridge" by Tristan Cazenave and Véronique Ventos proposes a search algorithm that improves on perfect information Monte Carlo search for some contracts of bridge. It repairs two known defects of perfect information Monte Carlo search, namely strategy fusion and non locality.

"Stabilized Nested Rollout Policy Adaptation" by Tristan Cazenave, Jean-Baptiste Sevestre, and Mathieu Toulemont presents a straightforward improvement of Nested Rollout Policy Adaptation (NRPA) that gives good results in SameGame, the Traveling

Salesman with Time Windows, and Expression Discovery. It stabilizes the learning of the policy by playing multiple playouts before adaptations at level one.

In "zoNNscan: A Boundary-Entropy Index for Zone Inspection of Neural Models", Adel Jaouen and Erwan Le Merrer introduce an index that is intended to inform the boundary uncertainty (in terms of the presence of other classes) around one given input data point. It is based on confidence entropy, and is implemented through Monte Carlo sampling in the multidimensional ball surrounding that input.

Reward shaping is often a hidden critical component of reinforcement learning. In "Ordinal Monte Carlo Tree Search", Tobias Joppen and Johannes Fürnkranz discuss the use of only a ranking information on rewards (as opposed to numerical values). This makes their proposed Monte Carlo tree search (MCTS) variant robust to poorly chosen reward values.

"Monte Carlo Game Solver" by Tristan Cazenave uses the information on move ordering collected during a Monte Carlo tree search to order moves in Alpha-Beta search. Various small games are solved with this Alpha-Beta variant equipped with Monte Carlo move ordering. Overall, it improves on Alpha-Beta with simple move ordering heuristics.

"Generalized Nested Rollout Policy Adaptation" by Tristan Cazenave demonstrates mathematically that Generalized Nested Rollout Policy Adaptation with a bias and a temperature is equivalent to Nested Rollout Policy Adaptation with initialization of the weights and a different learning rate. Using a bias is more convenient and more general than initializing weights as demonstrated for SameGame. Good results are also obtained for the Traveling Salesman Problem with Time Windows.

"Monte Carlo Inverse Folding" by Tristan Cazenave and Thomas Fournier addresses the design of RNA molecules with Monte Carlo search. They test various optimizations of NRPA on the RNA design problem and reach state-of-the-art results on the Eterna100 benchmark using general improvements of Monte Carlo search.

"Monte Carlo Graph Coloring" by Tristan Cazenave, Benjamin Negrevergne and Florian Sikora, deals with the difficult combinatorial optimization problem of assigning colors to vertices of a graph so that two adjacent vertices have different colors. They show that Monte Carlo search can be competitive for some of the problems of a standard problem suite.

"Enhancing Playout Policy Adaptation for General Game Playing" by Chiara Sironi, Tristan Cazenave and Mark Winands, compares various playout optimization strategies in the context of general game playing. Playout Policy Adaptation and MAST are compared as well as their enhanced versions using N-grams, discounting, and payoff driven policy updating.

Two other papers were presented at the workshop but are not included in the proceedings since they have already been published. They are "Trial-based Heuristic Tree Search for MDPs with Factored Action Spaces" by Florian Geisser, David Speck and Thomas Keller, and "Integrating Acting, Planning, and Learning in Hierarchical Operational Models" by Sunandita Patra, James Mason, Amit Kumar, Malik Ghallab, Paolo Traverso, and Dana Nau.

These proceedings would not have been produced without the help of many people. In particular, we would like to mention the authors and reviewers for their help. Moreover, the organizers of IJCAI 2020 contributed substantially by bringing the researchers together.

April 2021

Tristan Cazenave
Olivier Teytaud
Mark H. M. Winands

Organization

Program Chairs

Tristan Cazenave   Université Paris-Dauphine, France
Olivier Teytaud   FAIR Paris, France
Mark H. M. Winands   Maastricht University, The Netherlands

Program Committee

Yngvi Björnsson   Reykjavik University, Iceland
Bruno Bouzy   Nukkai, France
Cameron Browne   Maastricht University, The Netherlands
Tristan Cazenave   Université Paris-Dauphine, France
Stefan Edelkamp   King's College London, UK
Raluca Gaina   Queen Mary University of London, UK
Aurélien Garivier   ENS Lyon, France
Reijer Grimbergen   Tokyo University of Technology, Japan
Nicolas Jouandeau   University of Paris 8 Vincennes-Saint-Denis, France
Emilie Kaufmann   CNRS, France
Jakub Kowalski   University of Wrocław, Poland
Marc Lanctot   Google DeepMind, Canada
Jialin Liu   Southern University of Science and Technology, China
Martin Müller   University of Alberta, Canada
Andrzej Nagórko   University of Warsaw, Poland
Benjamin Negrevergne   Université Paris-Dauphine, France
Santiago Ontañón   Drexel University, USA
Diego Perez-Liebana   Queen Mary University of London, UK
Mike Preuss   Leiden University, The Netherlands
Thomas Runarsson   University of Iceland, Iceland
Abdallah Saffidine   University of New South Wales, Australia
Spyridon Samothrakis   University of Essex, UK
Chiara Sironi   Maastricht University, The Netherlands
Fabien Teytaud   Université Littoral Côte d'Opale, France
Olivier Teytaud   Facebook FAIR, France
Ruck Thawonmas   Ritsumeikan University, Japan
Jean-Noël Vittaut   Sorbonne University, France
Mark Winands   Maastricht University, The Netherlands
I-Chen Wu   National Chiao Tung University, Taiwan

Additional Reviewer

Maarten Schadd

Contents

The $\alpha\mu$ Search Algorithm for the Game of Bridge..................... 1
Tristan Cazenave and Véronique Ventos

Stabilized Nested Rollout Policy Adaptation........................ 17
Tristan Cazenave, Jean-Baptiste Sevestre, and Matthieu Toulemont

zoNNscan: A Boundary-Entropy Index for Zone Inspection
of Neural Models 31
Adel Jaouen and Erwan Le Merrer

Ordinal Monte Carlo Tree Search 39
Tobias Joppen and Johannes Fürnkranz

Monte Carlo Game Solver 56
Tristan Cazenave

Generalized Nested Rollout Policy Adaptation 71
Tristan Cazenave

Monte Carlo Inverse Folding 84
Tristan Cazenave and Thomas Fournier

Monte Carlo Graph Coloring 100
Tristan Cazenave, Benjamin Negrevergne, and Florian Sikora

Enhancing Playout Policy Adaptation for General Game Playing.......... 116
Chiara F. Sironi, Tristan Cazenave, and Mark H. M. Winands

Author Index 141

The $\alpha\mu$ Search Algorithm for the Game of Bridge

Tristan Cazenave[1(✉)] and Véronique Ventos[2]

[1] LAMSADE, Université Paris-Dauphine, PSL, CNRS, Paris, France
Tristan.Cazenave@dauphine.psl.eu
[2] NUKKAI, Paris, France
vventos@nukk.ai

Abstract. $\alpha\mu$ is an anytime heuristic search algorithm for incomplete information games that assumes perfect information for the opponents. $\alpha\mu$ addresses and if given enough time solves the strategy fusion and the non-locality problems encountered by Perfect Information Monte Carlo search (PIMC). Strategy fusion is due to PIMC playing different strategies in different worlds when it has to find a unique strategy for all the worlds. Non-locality is due to choosing locally optimal moves that are globally inferior. In this paper $\alpha\mu$ is applied to the game of Bridge and outperforms PIMC.

1 Introduction

As computer programs have reached superhuman at Go [18] and other two-player perfect information games like Chess and Shogi [17] starting from zero knowledge, some of the next challenges in games are imperfect information games such as Bridge or Poker. Multiplayer Poker has been solved very recently [1] while Computer Bridge programs are still not superhuman.

The state of the art for Computer Bridge is Perfect Information Monte Carlo search. It is a popular algorithm for imperfect information games. It was first proposed by Levy [12] for Bridge, and used in the popular program GIB [9]. PIMC can be used in other trick-taking card games such as Skat [2,11], Spades and Hearts [20]. The best Bridge and Skat programs use PIMC. Long analyzed the reasons why PIMC is successful in these games [14].

However PIMC plays sub-optimally due to two main problems: strategy fusion and non-locality. We will illustrate these problems in the second section. Frank and Basin [5] have proposed a heuristic algorithm to solve Bridge endgames that addresses the problems of strategy fusion and non-locality for late endgames. The algorithm we propose is an improvement over the algorithm of Frank and Basin since it solves exactly the endgames instead of heuristically and since it can also be used in any state even if the search would be too time consuming for the program to reach terminal states. Ginsberg has proposed to use a lattice and binary decision diagrams to improve the approach of Frank and Basin for solving Bridge endgames [9]. He states that he was generally able to

T. Cazenave et al. (Eds.): MCS 2020, CCIS 1379, pp. 1–16, 2021.
https://doi.org/10.1007/978-3-030-89453-5_1

solve 32 cards endings, but that the running times were increasing by two orders of magnitude as each additional card was added. $\alpha\mu$ is also able to solve Bridge endings but it can also give a heuristic answer at any time and for any number of cards and adding cards or searching deeper does not increase as much the running time.

Furtak has proposed recursive Monte Carlo search for Skat [7] to improve on PIMC but the algorithm does not give exact results in the endgame and does not solve the non-locality problem.

Other approaches to imperfect information games are Information Set Monte Carlo Tree Search [3], counterfactual regret minimization [22], and Exploitability Descent [13].

$\alpha\mu$ searches with partial orders. It is related to partial order bounding [16] and to opponent modeling in card games [19]. However our algorithm is different from these algorithms since it searches over vectors only composed of 0 and 1 and uses different backups for sets of vectors at Max and Min nodes as well as probabilities of winning.

The contributions of the paper are:

1. An anytime heuristic search algorithm that assumes Min players have perfect information and that improves on PIMC and previous related search algorithms.
2. An anytime solution to the strategy fusion problem of PIMC that solves the strategy fusion problem when given enough time.
3. An anytime solution to the non-locality problem of PIMC using Pareto fronts of vectors representing the outcomes for the different possible worlds. It also converges given enough time.
4. A search algorithm with Pareto fronts.
5. The description of the early and root cuts that speed up the search.
6. Adaptation of a transposition table to the algorithm so as to improve the search speed using iterative deepening.
7. Experimental results for the game of Bridge.

The paper is organized as follows: the second section deals with Bridge and Computer Bridge. The third section defines vectors of outcomes and Pareto fronts. The fourth section details the $\alpha\mu$ algorithm. The fifth section gives experimental results.

2   Bridge and Computer Bridge

2.1   Bridge in Short

The interested reader can refer for instance to [15] for a more complete presentation of the game of Bridge. Bridge is a trick-taking card game opposing four players (denoted by West, North, East and South or W, N, E, S) divided in two partnerships (East-West and North-South). A standard 52 card pack is shuffled and each player receives a hand of 13 cards that is only visible to him. A Bridge game is divided into two major playing phases: the bidding phase (out of the scope of the paper) and the card play. The goal of the bidding phase is to reach a contract which determines the minimum number of tricks the pair commits to win during the card play, either with no trump (NT) or with a determined suit as trump. During the card play, the goal is to fulfill (for the declarer) or to defeat (for the defenders) the contract reached during the bidding phase. Let us assume that South is the agent who plays the game (i.e. the declarer). Player on the left of the declarer (W) exposes the first card of the game. The declarer's partner (N called Dummy) then lays his cards face up on the table. When playing in a NT contract, there is only one simple rule: each player is required to follow suit if possible and can play any card of the suit. When the four players have played a card, the player who played the highest-ranked card in the suit (2 < 3 < ... < 10 < J < Q < K < A) wins the trick and he will be on lead at the following trick. The game is over when all the cards have been played, In the following (including in our experiments), we assume that the pair North-South (NS) reached the contract of 3NT. In this case, if NS wins nine or more tricks the game is won otherwise it is lost[1].

2.2 Computer Bridge

A Double Dummy Solver (DDS) is a solver for complete information Bridge. A very efficient Double Dummy Solver (DDS) has been written by Bo Haglund [10]. In our experiments we use it to evaluate double dummy hands. It makes use of partition search [8] among many other optimizations to improve the solving speed of the $\alpha\beta$.

PIMC is the state of the art of Computer Bridge, it is used for example in GIB [9] and in WBRIDGE5 [21] the former computer world champion.

The PIMC algorithm is given in Algorithm 1. In this algorithm S is the set of possible worlds and *allMoves* is the set of moves to be evaluated. The play function plays a move in a possible world and returns the corresponding state. All the possible worlds have the same hand for the player to play, so all the moves for the player to play are legal in all the possible worlds. The DDS function evaluates the state using a double dummy solver. The DDS sends back the maximum number of tricks the player to play can win. If the number of tricks already won by the declarer plus the number of tricks that the declarer can win returned by DDS is greater than or equal to the contract the world is evaluated to 1, else to 0.

---

[1] It is an acceptable simplification of the real scoring of Bridge. At Bridge, the declarer has to make six more tricks than the number in his contract.

Algorithm 1. The PIMC algorithm.

1: Function $PIMC$ ($allMoves, S$)
2:    **for** $move \in allMoves$ **do**
3:        $score[move] \leftarrow 0$
4:        **for** $w \in S$ **do**
5:            $s \leftarrow$ play $(move, w)$
6:            $score[move] \leftarrow score[move] +$ DDS (s)
7:        **end for**
8:    **end for**
9:    **return** $argmax_{move}(score[move])$

PIMC accumulates the payoff of strategies related to different worlds. This process leads to an optimistic evaluation since in reality a specific strategy has to be chosen. This problem is known as strategy fusion [4]. The reason why PIMC is optimistic is that it can adapt its strategy to each world since it has perfect information. In the real game the player does not know the real world and cannot adapt its strategy, it has to choose a strategy working in all possible worlds.

♠KJT7
♡AKQ
♢AKQ
♣xxx

♠A986
♡xxx
♢xxx
♣AKQ

Fig. 1. Example of a Bridge hand illustrating strategy fusion.

Figure 1 is a hand from [9] which illustrates the strategy fusion problem. PIMC finds that the declarer always makes all of the four tricks at Spades when it has only 50% chances of making them since it has to finesse the Queen. The declarer does not know where is the Queen, so it has to bet where she is and for example play the Jack for the dummy hoping she is not in East hand.

Strategy fusion arises because PIMC can play different cards in different worlds whereas it should play the same cards in all the worlds since it cannot distinguish between worlds. Frank and Basin solve this problem with an algorithm they call Vector Minimaxing [5] that plays the same cards for the Max player in all the worlds.

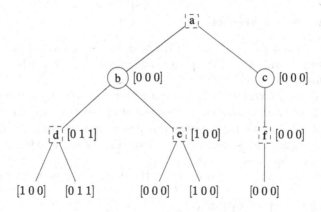

Fig. 2. Example of a tree with three worlds illustrating non-locality.

The fact that a move is optimal at a node but not optimal considering the whole search tree leads to the problem of non-locality. From an algorithmic point of view non-locality can be explained using Fig. 2 from [6]. It illustrates non-locality when searching with strategy fusion for Max and perfect information for Min. As usual the Max nodes are squares and the Min nodes are circles. The Max nodes are dashed since they represent information sets for Max. Max does not know in which of the three worlds he is playing whereas Min can distinguish the worlds and thus chooses actions in a different way for each world. The leaves give the result of the game in the three possible worlds. For example the move to the left from node d reaches a state labeled [1 0 0] which means that the game is won in world 1 (hence the 1 in the first position), lost in world 2 (hence the 0 in the second position) and also lost in world 3 (hence the 0 in the third position). The vectors near the internal nodes give the values that are backed up by the strategy fusion for Max and perfect information for Min algorithm. We can see that each Max node is evaluated by choosing the move that gives the maximum average outcome. For example at node d there are two moves, the left one leads to [1 0 0] and therefore has an average of $\frac{1}{3}$ whereas the right one leads to [0 1 1] and has an average of $\frac{2}{3}$. So node d backs up [0 1 1]. However it is not globally optimal. If instead of choosing the right move at node d it chooses the left move it backs up [1 0 0] and then the b node would have been evaluated better also with [1 0 0]. It illustrates that choosing the local optimum at node d prevents from finding the real optimum at node b. At Min nodes the algorithm chooses for each world the minimal outcome over all children since it can choose the move it prefers most in each different world.

3   Vectors of Outcomes and Pareto Fronts

In this section we define Vectors and Pareto fronts that are used by the algorithms in the next section.

3.1 Definitions for Vectors

Given n different possible worlds, a vector of size n keeps the status of the game for each possible world. A zero (resp. one) at index i means that the game is lost (resp. won) for world number i.

Associated to each vector there is another vector of booleans indicating which worlds among the n are possible in the current state. At the root of the search all worlds are possible but when an opponent makes a move, the move is usually only valid in some of the worlds, the associated vector is then updated by changing from true to false for these worlds.

The associated vector is used to define the domination between two vectors:

v1 $\geq$ v2 iff $\forall i \in [1,n]$, v1[i] $\geq$ v2[i]
v1 dominates v2 iff they have the same associated worlds,
v1 $\geq$ v2 and $\exists i \in [1,n]$ such that v1[i] $>$ v2[i].

The score of a vector is the average among all possible worlds of the values contained in the vector.

3.2 Pareto Front

A Pareto front is a set of vectors. It maintains the set of vectors that are not dominated by other vectors. When a new vector is a candidate for insertion in the front the first thing to verify is whether the candidate vector is dominated by a vector in the front or equal to another vector in the front. If it is the case the candidate vector is not inserted and the front stays the same. If the candidate vector is not dominated it is inserted in the front and all the vectors in the front that are dominated by the candidate vector are removed.

For example consider the Pareto front $\{[1\ 0\ 0], [0\ 1\ 1]\}$. If the vector $[0\ 0\ 1]$ is a candidate for entering the front, then the front stays unchanged since $[0\ 0\ 1]$ is dominated by $[0\ 1\ 1]$. If we add the vector $[1\ 1\ 0]$ then the vector $[1\ 0\ 0]$ is removed from the front since it is dominated by $[1\ 1\ 0]$, and then $[1\ 1\ 0]$ is inserted in the front. The new front becomes $\{[1\ 1\ 0], [0\ 1\ 1]\}$.

It is useful to compare Pareto fronts. A Pareto front P_1 dominates or is equal to a Pareto front P_2 iff $\forall v \in P_2$, $\exists v' \in P_1$ such that (v' dominates v) or v' = v.

4   The $\alpha\mu$ Algorithm

In this section we first explain how the algorithm deals with strategy fusion and non-locality, we then give some optimizations and we eventually explain the details of the search algorithm.

4.1   Search with Strategy Fusion

Let us assume that the defense knows the cards of the declarer and that the declarer optimizes against all possible states that correspond to his information. The score of a move for the declarer is the highest score of all vectors in the Pareto front of the move. At a Max node the declarer computes after each move the union of the Pareto fronts of all the moves that have been tried so far. Min has knowledge of the declarer cards so in each world he takes the move that minimizes the result of Max. The code for Min and Max nodes is given in Algorithm 2. $\alpha\mu$ is a generalization of PIMC since a search with a depth of one is PIMC.

The parameter M controls the number of Max moves, when $M = 0$ the algorithm reaches a leaf and each remaining possible world is evaluated with a double dummy search. The stop function also stops the search if the game is already won no matter what is played after. The parameter *state* contains the current state where all the moves before have been played and which does not contain the hidden information. The parameter *Worlds* contains the set of all possible worlds compatible with the moves already played. The transposition table contains the Pareto front and the best move found by the previous search of a state. If the state has not yet been searched the Pareto front is initialized as the empty set and the best move is not set. If at Min node, the set of all possible moves in all possible worlds is calculated (lines 13–16). At each played move the list of possible worlds is updated and a recursive call performed. The Pareto front resulting from the recursive call is then combined with the overall front (lines 18–23). We will explain later the min algorithm. Similar operations are performed for a Max node except that the combination with the overall front is then done with the max algorithm (lines 27–45). We explain the max algorithm in the next subsection. The optimizations and detailed explanations of the algorithm are given in Subsects. 4.3 and 4.4.

4.2 Dealing with Non-locality

Max Nodes. At Max nodes each possible move returns a Pareto front. The overall Pareto front is the union of all the Pareto fronts of the moves. The idea is to keep all the possible options for Max, i.e. Max has the choice between all the vectors of the overall Pareto front.

Min Nodes. The Min players can choose different moves in different possible worlds. So they take the minimum outcome over all the possible moves for a possible world. So when they can choose between two vectors they take for each index the minimum between the two values at this index of the two vectors.

Now when Min moves lead to Pareto fronts, the Max player can choose any member of the Pareto front. For two possible moves of Min, the Max player can also choose any combination of a vector in the Pareto front of the first move and of a vector in the Pareto front of the second move. In order to build the Pareto front at a Min node we therefore have to compute all the combinations of the vectors in the Pareto fronts of all the Min moves. For each combination the minimum outcome is kept so as to produce a unique vector. Then this vector is inserted in the Pareto front of the Min node.

An example of the product of Pareto fronts is given in Fig. 3. We can see in the figure that the left move for Min at node a leads to a Max node b with two moves. The Pareto front of this Max node is the union of the two vectors at the leaves: $\{[0\ 1\ 1], [1\ 1\ 0]\}$. The right move for Min leads to a Max node c with three possible moves. When adding the vectors to the Pareto front of the Max node c, the algorithm sees that $[1\ 0\ 0]$ is dominated by $[1\ 0\ 1]$ and therefore does not add it to the Pareto front at node c. So the resulting Pareto front for the Max node c is $\{[1\ 1\ 0], [1\ 0\ 1]\}$. Now to compute the Pareto front for the root Min node we perform the product of the two reduced Pareto fronts of the children Max nodes and it gives: $\{[0\ 1\ 0], [0\ 0\ 1], [1\ 1\ 0], [1\ 0\ 0]\}$. We then reduce the Pareto front of the Min node and remove $[0\ 1\ 0]$ which is dominated by $[1\ 1\ 0]$ and also remove $[1\ 0\ 0]$ which is also dominated by $[1\ 1\ 0]$. Therefore the resulting Pareto front for the root Min node is $\{[0\ 0\ 1], [1\ 1\ 0]\}$.

We can also explain the behavior at Min nodes on the non-locality example of Fig. 2. The Pareto front at Max node d is $\{[1\ 0\ 0], [0\ 1\ 1]\}$. The Pareto front at Max node e is $\{[0\ 0\ 0], [1\ 0\ 0]\}$. It is reduced to $\{[1\ 0\ 0]\}$ since $[0\ 0\ 0]$ is dominated. Now at node b the product of the Pareto fronts at nodes d and e gives $\{[1\ 0\ 0], [0\ 0\ 0]\}$ which is also reduced to $\{[1\ 0\ 0]\}$. The Max player can now see that the b node is better than the c node, it was not the case for the strategy fusion algorithm without Pareto fronts.

4.3 Optimizations

In this section we explain how to speedup search.

Skipping Min Nodes. A one ply search at a Min node will always give the same result as the Pareto front at that node since the Double Dummy Solver has already searched all worlds with an $\alpha\beta$. The Min player can choose the move for each world and therefore will have the same result as the $\alpha\beta$ for each world.

Algorithm 2. The $\alpha\mu$ search algorithm with cuts and transposition table.

```
 1: Function αμ (state, M, Worlds, α)
 2:     if stop(state, M, Worlds, result) then
 3:         update the transposition table
 4:         return result
 5:     end if
 6:     t ← entry of state in the transposition table
 7:     if Min node then
 8:         mini ← ∅
 9:         if t.front ≤ α then
10:             return mini
11:         end if
12:         allMoves ← ∅
13:         for w ∈ Worlds do
14:             l ← legalMoves (w)
15:             allMoves = allMoves ∪ l
16:         end for
17:         move t.move in front of allMoves
18:         for move ∈ allMoves do
19:             s ← play (move, state)
20:             W₁ ← {w ∈ Worlds : move ∈ w}
21:             f ← αμ (s, M, W₁, ∅)
22:             mini ← min(mini, f)
23:         end for
24:         update the transposition table
25:         return mini
26:     else
27:         front ← ∅
28:         allMoves ← ∅
29:         for w ∈ Worlds do
30:             l ← legalMoves (w)
31:             allMoves = allMoves ∪ l
32:         end for
33:         move t.move in front of allMoves
34:         for move ∈ allMoves do
35:             s ← play (move, state)
36:             W₁ ← {w ∈ Worlds : move ∈ w}
37:             f ← αμ (s, M − 1, W₁, front)
38:             front ← max(front, f)
39:             if root node then
40:                 if μ(front) = μ of previous search then
41:                     break
42:                 end if
43:             end if
44:         end for
45:         update the transposition table
46:         return front
47:     end if
```

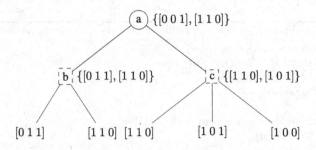

Fig. 3. Product of Pareto fronts at Min nodes.

This is why we only keep the number M of Max moves to be played in the search. The search will never stop after a Min move since recursive calls at Min node do not decrease M. This is intended since the results of the search after a Min move are the same as before the Min move.

Iterative Deepening and Transposition Table. Iterative deepening starts with one Max move and increases the number of Max moves at every iteration. The number of Max moves is the number of Max nodes that have been traversed before reaching the current state. The results of previous searches for all the nodes searched are stored in a transposition table.

An entry in the transposition table contains the Pareto front of the previous search at this node and the best move found by the search. The best move is the move with the greatest probability.

When a search is finished at a node, the entry in the transposition table for this node is updated with the new Pareto front and the new best move.

Comparing Pareto Fronts at Min Nodes. When a Pareto front P_1 dominates another Pareto front P_2 it is safe to ignore the move associated to P_2 since it adds no options to P_1. If it is true for the current front P_2 at a Min node it will also be true when searching more this Min node since P_2 can only be reduced to an even more dominated Pareto front by more search at a Min node.

The Early Cut. If a Pareto front at a Min node is dominated by the Pareto front of the upper Max node it can safely be cut since the evaluation is optimistic for the Max player. The Max player cannot get a better evaluation by searching more under the Min node and it will always be cut whatever the search below the node returns since the search below will return a Pareto front smaller or equal to the current Pareto front. It comes from the observation that a world lost at a node can never become won.

Figure 4 gives an example of an early cut at a Min node. The root node a is a Max node, the first move played at a returned $\{[1\ 1\ 0], [0\ 1\ 1]\}$ which is backed up at node a. The second move is then tried leading to node c and the initial Pareto front calculated with double dummy searches at node c is $[1\ 1\ 0]$. It is dominated by the Pareto front of node a so node c can be cut.

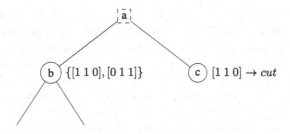

Fig. 4. Example of an early cut at node c.

The Root Cut. If a move at the root of $\alpha\mu$ for M Max moves gives the same probability of winning than the best move of the previous iteration of iterative deepening for $M - 1$ Max moves, the search can safely be stopped since it is not possible to find a better move. A deeper search will always return a worse probability than the previous search because of strategy fusion. Therefore if the probability is equal to the one of the best move of the previous shallower search the probability cannot be improved and a better move cannot be found so it is safe to cut.

4.4   Detailed Algorithm

$\alpha\mu$ with transposition table and cuts is a search algorithm using Pareto fronts as evaluations and bounds. The algorithm is given in Algorithm 2.

The evaluation of a state at a leaf node is the double dummy evaluation for each possible world. An evaluation for a world is 0 if the game is lost for the Max player and 1 if the game is won for the Max player (lines 2–5).

The algorithm starts with getting the entry t of *state* in the transposition table (line 6). The entry contains the last Pareto front found for this state and the best move found for this state, i.e. the move associated to the best average.

If the state is associated to a Min node, i.e. a Min player is to play, the algorithm starts to get the previously calculated Pareto front from the transposition table (line 8). Then it looks for an early cut (lines 9–11). If the node is not cut it computes the set of all possible moves over all the valid worlds (lines 12–16). It then moves the move of the transposition table in front of the possible moves (line 17). After that it tries all possible moves (line 18). For each possible move it computes the set W_1 of worlds still valid after the move and recursively calls $\alpha\mu$ (lines 19–21) . The parameters of the recursive call are s, the current state, M the number of Max moves to go which is unchanged since we just played a Min move, W_1 the set of valid worlds after *move*, and an empty set for *alpha* to avoid deeper cuts. The front returned by the recursive call is then combined to the current front using the min function (line 22). When the search is finished it updates the transposition table and returns the *mini* Pareto front (lines 24–25).

If the state is associated to a Max node it initializes the resulting front with an empty set (line 27). Then as in the Min nodes it computes the set of all possible moves and moves the transposition table move in front of all the possible moves (lines 28–32). Then it tries all the moves and for each move computes the new set W_1 of valid worlds and recursively calls $\alpha\mu$ with $M-1$ since a Max move has just been played and $front$ as $alpha$ since a cut can happen below when the move does not improve $front$ (lines 33–36). The resulting front f is combined with front with the max function (line 37). If the score of the best move ($\mu(front)$) is equal to the score of the best move of the previous search and the node is the root node then a Root cut is performed (lines 38–42). When the search is finished the transposition table is updated and $front$ is returned (lines 44–45).

The search with strategy fusion is always more difficult for the Max player than the double dummy search where the Max player can choose different moves in the different possible worlds for the same state. Therefore if a double dummy search returns a loss in a possible world, it is sure that the search with $\alpha\mu$ will also return a loss for this world.

If the search is performed until terminal nodes and all possible worlds are considered then $\alpha\mu$ solves the strategy fusion and the non locality problems for the game where the defense has perfect information.

If the search is stopped before terminal nodes and not all possible worlds are considered then $\alpha\mu$ is a heuristic search algorithm. The algorithm is named $\alpha\mu$ since it maximizes the mean and uses an α bound.

5   Experimental Results

In our experiments we fix the bid so as to concentrate on the evaluation of the card play. We use duplicate scoring. It means that the different evaluated programs will play the same initial deals against the same opponents. When $\alpha\mu$ is the declarer it will play against two PIMC as the defense. In the following we note $\alpha\mu(1)$ for $\alpha\mu$ with one Max move and $\alpha\mu(3)$ for $\alpha\mu$ with three Max moves. In order to compare $\alpha\mu$ as a declarer to PIMC as a declarer we compare $\alpha\mu(3)$ as a declarer to $\alpha\mu(1)$ as a declarer since $\alpha\mu(1)$ is PIMC. In most of the experiments $\alpha\mu$ with 20 possible worlds plays against PIMC with 20 possible worlds as the defense. All results are computed playing the same initial deals with the same seed for each deal, meaning that as long as the card played are the same, the generated possible worlds for $\alpha\mu(1)$ and $\alpha\mu(3)$ are the same. The Pareto fronts stay small in practice. For Bridge it is very often the case with random deals that a world champion and a weak player score the same. It is more informative to compute statistics on deals with different results. Challenges in Bridge are played with carefully selected deals not random ones.

Table 1 gives the results for games that have a different result for $\alpha\mu(1)$ and $\alpha\mu(3)$. We only keep the games that are either won for $\alpha\mu(1)$ and lost for $\alpha\mu(3)$ or lost for $\alpha\mu(1)$ and won for $\alpha\mu(3)$. We stop the experiment when 200 such games have been played. The winrate of $\alpha\mu(3)$ is significantly greater than $\alpha\mu(1)$. For example with 32 cards, $\alpha\mu(3)$ wins 62% of the time with a standard deviation of 3.4%.

PIMC (i.e. $\alpha\mu(1)$) is already a very strong player as a declarer so improving on it even slightly is difficult. We can conclude that looking three Max moves ahead is beneficial and that $\alpha\mu(3)$ improves on PIMC.

In practice the number of vectors in the Pareto fronts always stayed very small in the games it played.

Table 1. Comparison of $\alpha\mu(3)$ with PIMC for games that have a different result. 3NT contract.

Cards	M	Worlds	Games	Winrate	σ
32	1	20	200	0.380	0.034
32	3	20	200	0.620	0.034
36	1	20	200	0.455	0.035
36	3	20	200	0.545	0.035
40	1	20	200	0.450	0.035
40	3	20	200	0.550	0.035
52	1	20	200	0.420	0.035
52	3	20	200	0.580	0.035

Table 2 gives the percentage of games played by $\alpha\mu(3)$ that are different from the games played by $\alpha\mu(1)$ it also gives the percentage of the total number of games that have different results. In practice when generating the initial deals randomly, many deals are not interesting: they are either easily won or completely lost. The two algorithms play the same cards on these deals as there is no interesting alternative. This is the reason why we only keep the games where the algorithms have different results so as to evaluate the difference between the two algorithms.

Table 2. Percentage of different games for $\alpha\mu(3)$ for the 3NT contract.

Cards	Worlds	Different	Different Results
32	20	10.4%	2.4%
36	20	16.4%	3.8%
40	20	23.6%	5.8%
52	20	40.9%	10.6%

We now compare the times to play moves with and without Transposition Tables and cuts. Table 3 gives the average time per move of different configurations of $\alpha\mu$ playing entire games. The initial deals used for this experiment are the first 100 initial deals of the previous experiment for a total of 2 600 cards played.

TT means Transposition Table, R means Root Cut, E means Early Cut. We can observe that a Transposition Table associated to cuts improves the search time. For $M = 1$ the search time is 0.159 s. For $M = 3$ without transposition table and cuts the average search time per move is 55 s. When using a transposition table associated to early and root cuts it goes down to 3 s.

Table 3. Comparison of the average time per move of different configurations of $\alpha\mu$ on deals with 52 cards for the 3NT contract.

Cards	M	Worlds	TT	R	E	Time
52	1	20				0.118
52	2	20	n	n	n	1.054
52	2	20	y	y	n	0.512
52	2	20	y	n	y	0.503
52	2	20	y	y	y	0.433
52	3	20	n	n	n	10.276
52	3	20	y	y	n	3.891
52	3	20	y	n	y	1.950
52	3	20	y	y	y	1.176

We also made experiments with the 7NT contract. In this contract the declarer has to win all the tricks. When a trick is won by the defense the search can stop as the contract is lost. Table 4 gives the results for 2, 3 and 4 Max moves and 20 and 40 possible worlds. We compute statistics on the games that have different results for PIMC and $\alpha\mu$ out of the 10 000 games played for each experiment. For example the first line gives the comparison of $\alpha\mu$ with two Max moves and 20 worlds against PIMC with 20 worlds for 10 000 games, 283 of these 10 000 games give different outcomes for the two algorithms, 64.3% of these 283 games are won by $\alpha\mu$ and lost by PIMC. For the 40 worlds experiments $\alpha\mu$ with 40 worlds is compared to PIMC with 40 worlds.

Table 4. Comparison of $\alpha\mu$ versus PIMC for the 7NT contract, playing 10 000 games.

Cards	M	Worlds	$\neq$ results	Winrate	σ
52	2	20	283	0.643	0.0285
52	3	20	333	0.673	0.0257
52	4	20	374	0.679	0.0241
52	2	40	324	0.630	0.0268
52	3	40	347	0.637	0.0258
52	4	40	368	0.655	0.0248

6 Conclusion and Future Work

We presented $\alpha\mu$, which is a heuristic search algorithm for incomplete information games. In order to highlight its advantages, we tested $\alpha\mu$ on the card play of Bridge, which is known to be difficult for classical search algorithms such as PIMC according to the strategy fusion and non-locality problems. To solve the non-locality problem $\alpha\mu$ uses Pareto fronts as evaluations of states and combines them in an original way at Min and Max nodes. To solve the strategy fusion problem it plays the same moves in all the valid worlds during search. Experimental results for the 3NT contract and even more for the 7NT contract show it significantly improves on PIMC, which is a breakthrough in the field of Computer Bridge.

We also presented the use of a transposition table as well as the early and the root cut for $\alpha\mu$. When searching three Max moves ahead it enables the search to be faster while returning the same move as the longer search without the optimizations.

In future work we expect to use partition Search with $\alpha\mu$ and to speed it up with other cuts and optimizations. We plan to carry out new experiments with $\alpha\mu$ for the defense, and with real scores instead of only win/loss. We also plan to parallelize the algorithm. The algorithm is easy to parallelize strongly, for example parallelizing the DDS calls at the leaves or parallelizing the search for the different moves at the root. The sequential times are not indicative of the time limits once parallelized. Finally, a promising approach for improving $\alpha\mu$ is to make inferences linked to the strategy of the opponents in order to reduce the set of possible worlds.

Acknowledgment. Thanks to Alexis Rimbaud for explaining me how to use the solver of Bo Haglund and to Bo Haglund for his Double Dummy Solver.

References

1. Brown, N., Sandholm, T.: Superhuman AI for multiplayer poker. Science **365**(6456), 885–890 (2019)
2. Buro, M., Long, J.R., Furtak, T., Sturtevant, N.: Improving state evaluation, inference, and search in trick-based card games. In: Twenty-First International Joint Conference on Artificial Intelligence (2009)
3. Cowling, P.I., Powley, E.J., Whitehouse, D.: Information set Monte Carlo tree search. IEEE Trans. Comput. Intell. AI Games **4**(2), 120–143 (2012)
4. Frank, I., Basin, D.: Search in games with incomplete information: a case study using bridge card play. Artif. Intell. **100**(1–2), 87–123 (1998)
5. Frank, I., Basin, D.: A theoretical and empirical investigation of search in imperfect information games. Theoret. Comput. Sci. **252**(1–2), 217–256 (2001)
6. Frank, I., Basin, D.A., Matsubara, H.: Finding optimal strategies for imperfect information games. In: AAAI/IAAI, pp. 500–507 (1998)
7. Furtak, T., Buro, M.: Recursive Monte Carlo search for imperfect information games. In: 2013 IEEE Conference on Computational Intelligence in Games (CIG), pp. 1–8. IEEE (2013)

8. Ginsberg, M.L.: Partition search. In: Proceedings of the Thirteenth National Conference on Artificial Intelligence and Eighth Innovative Applications of Artificial Intelligence Conference, AAAI 1996, IAAI 1996, Portland, Oregon, USA, 4–8 August 1996, vol. 1, pp. 228–233 (1996). http://www.aaai.org/Library/AAAI/1996/aaai96-034.php

9. Ginsberg, M.L.: GIB: imperfect information in a computationally challenging game. J. Artif. Intell. Res. **14**, 303–358 (2001). https://doi.org/10.1613/jair.820

10. Haglund, B.: Search algorithms for a bridge double dummy solver. Technical report (2010)

11. Kupferschmid, S., Helmert, M.: A Skat player based on Monte-Carlo simulation. In: van den Herik, H.J., Ciancarini, P., Donkers, H.H.L.M.J. (eds.) CG 2006. LNCS, vol. 4630, pp. 135–147. Springer, Heidelberg (2007). https://doi.org/10.1007/978-3-540-75538-8_12

12. Levy, D.N.: The million pound bridge program. In: Heuristic Programming in Artificial Intelligence The First Computer Olympiad, pp. 95–103 (1989)

13. Lockhart, E., et al.: Computing approximate equilibria in sequential adversarial games by exploitability descent. arXiv preprint arXiv:1903.05614 (2019)

14. Long, J.R., Sturtevant, N.R., Buro, M., Furtak, T.: Understanding the success of perfect information Monte Carlo sampling in game tree search. In: Twenty-Fourth AAAI Conference on Artificial Intelligence (2010)

15. Mahmood, Z., Grant, A., Sharif, O.: Bridge for Beginners: A Complete Course. Pavilion Books (2014)

16. Müller, M.: Partial order bounding: a new approach to evaluation in game tree search. Artif. Intell. **129**(1–2), 279–311 (2001)

17. Silver, D., et al.: A general reinforcement learning algorithm that masters chess, shogi, and go through self-play. Science **362**(6419), 1140–1144 (2018)

18. Silver, D., et al.: Mastering the game of go without human knowledge. Nature **550**(7676), 354 (2017)

19. Sturtevant, N., Zinkevich, M., Bowling, M.: Prob-max^ n: Playing n-player games with opponent models. In: AAAI, vol. 6, pp. 1057–1063 (2006)

20. Sturtevant, N.R., White, A.M.: Feature construction for reinforcement learning in hearts. In: van den Herik, H.J., Ciancarini, P., Donkers, H.H.L.M.J. (eds.) CG 2006. LNCS, vol. 4630, pp. 122–134. Springer, Heidelberg (2007). https://doi.org/10.1007/978-3-540-75538-8_11

21. Ventos, V., Costel, Y., Teytaud, O., Ventos, S.T.: Boosting a bridge artificial intelligence. In: 2017 IEEE 29th International Conference on Tools with Artificial Intelligence (ICTAI), pp. 1280–1287. IEEE (2017)

22. Zinkevich, M., Johanson, M., Bowling, M., Piccione, C.: Regret minimization in games with incomplete information. In: Advances in Neural Information Processing Systems, pp. 1729–1736 (2008)

Stabilized Nested Rollout Policy Adaptation

Tristan Cazenave[1]($\boxtimes$), Jean-Baptiste Sevestre[2], and Matthieu Toulemont[3]

[1] LAMSADE, Université Paris-Dauphine, PSL, CNRS, Paris, France
`Tristan.Cazenave@dauphine.psl.eu`
[2] InstaDeep, Paris, France
`jb.sevestre@instadeep.com`
[3] PhotoRoom, Paris, France
`matthieu.toulemont@ponts.org`

Abstract. Nested Rollout Policy Adaptation (NRPA) is a Monte Carlo search algorithm for single player games. In this paper we propose to modify NRPA in order to improve the stability of the algorithm. Experiments show it improves the algorithm for different application domains: SameGame, Traveling Salesman with Time Windows and Expression Discovery.

1 Introduction

Monte Carlo Tree Search (MCTS) has been successfully applied to many games and problems [3].

Nested Monte Carlo Search (NMCS) [4] is an algorithm that works well for puzzles and optimization problems. It biases its playouts using lower level playouts. At level zero NMCS adopts a uniform random playout policy. Online learning of playout strategies combined with NMCS has given good results on optimization problems [27]. Other applications of NMCS include Single Player General Game Playing [20], Cooperative Pathfinding [1], Software testing [25], heuristic Model-Checking [26], the Pancake problem [2], Games [10] and the RNA inverse folding problem [23].

Online learning of a playout policy in the context of nested searches has been further developed for puzzles and optimization with Nested Rollout Policy Adaptation (NRPA) [28]. NRPA has found new world records in Morpion Solitaire and crosswords puzzles. Stefan Edelkamp and co-workers have applied the NRPA algorithm to multiple problems. They have optimized the algorithm for the Traveling Salesman with Time Windows (TSPTW) problem [11,12]. Other applications deal with 3D Packing with Object Orientation [14], the physical traveling salesman problem [15], the Multiple Sequence Alignment problem [16] or Logistics [13]. The principle of NRPA is to adapt the playout policy so as to learn the best sequence of moves found so far at each level. Unfortunately, this mechanism only samples each policy once at the lowest level which may lead to a misclassification of a good policy (one that improves the best score) as a bad one. To solve this issue, we propose a simple, yet effective modification of

T. Cazenave et al. (Eds.): MCS 2020, CCIS 1379, pp. 17–30, 2021.
https://doi.org/10.1007/978-3-030-89453-5_2

the NRPA Algorithm, which we name Stabilized NRPA.By sampling each policy multiple times at the lowest level we show that this new NRPA is stabilized and converges faster.

We now give the outline of the paper. The second section describes NRPA. The third section explains Stabilized NRPA. The fourth section describes the problems used for the experiments. The fifth section gives experimental results for these problems. The sixth section outlines further work and the last section concludes.

2   NRPA

Nested Policy Rollout Adaptation is an algorithm introduced by Chris Rosin [28] that achieves state-of-the-art performance on problems such as Morpion Solitaire.

This algorithm has two major components: An adaptive rollout policy, and a nested structure, shown in Fig. 1.

The adaptive rollout policy is a policy parameterized by weights on each action. During the playout phase, action is sampled according to this weights. The Playout Algorithm is given in Algorithm 1. It uses Gibbs sampling, each move is associated to a weight. A move is coded as an integer that gives the index of its weight in the policy array of floats. The algorithm starts with initializing the sequence of moves that it will play (line 2). Then it performs a loop until it reaches a terminal states (lines 3–6). At each step of the playout it calculates the sum of all the exponentials of the weights of the possible moves (lines 7–10) and chooses a move proportionally to its probability given by the softmax function (line 11). Then it plays the chosen move and adds it to the sequence of moves (lines 12–13).

Then, the policy is adapted on the best current sequence found, by increasing the weight of the best actions. The Adapt Algorithm is given in Algorithm 2. For all the states of the sequence passed as a parameter it adds α to the weight of the move of the sequence (lines 3–5). Then it reduces all the moves proportionally to α times the probability of playing the move so as to keep a sum of all probabilities equal to one (lines 6–12).

The nested structure was introduced by Tristan Cazenave [4]. This method helps the algorithm to converge towards better and better sequences. In NRPA, each nested level takes as input a policy, and returns a sequence. Inside the level, the algorithm makes many recursive calls to lower levels, providing weights, getting sequences and adapting the weights on those sequences. In the end, the algorithm returns the best sequence found in that level. At the lowest level, the algorithm simply makes a rollout.

The NRPA algorithm is given in Algorithm 3. At level zero it simply performs a playout (lines 2–3). At greater levels it performs N iterations and for each iteration it calls itself recursively to get a score and a sequence (lines 4–7). If it finds a new best sequence for the level it keeps it as the best sequence (lines 8–11). Then it adapts the policy using the best sequence found so far at the current level (line 12).

NRPA balances exploitation by adapting the probabilities of playing moves toward the best sequence of the level, and exploration by using Gibbs sampling at the lowest level. It is a general algorithm that has proven to work well for many optimization problems.

3 Stabilized NRPA

In this section we explain Stabilized NRPA and its potential for being parallelized.

3.1 Better Convergence of NRPA

In NRPA algorithm, an evaluation problem may occur.

Imagine that we have a policy that has good performance, but unfortunately the sequence generated by this policy at level 0 is bad (i.e. the sequence has a bad score comparing to the usual policy performance). This sequence is up to level 1 and is ignored since it is worse than the best sequence of level 1. The policy is adapted on the best sequence of level 1, pushing slightly the next rollouts toward the best sequence of level 1, making the policy more deterministic, making it less exploratory and less likely to find a new best sequence. This bad behavior could be propagated to the upper level, for the same reasons.

Algorithm 1. The Playout algorithm

1: Playout (*state*, *policy*)
2:    *sequence* ← []
3:    **while** true **do**
4:        **if** *state* is terminal **then**
5:            **return** (score (*state*), *sequence*)
6:        **end if**
7:        $z \leftarrow 0.0$
8:        **for** m in possible moves for *state* **do**
9:            $z \leftarrow z + \exp (policy \, [\text{code}(m)])$
10:       **end for**
11:       choose a *move* with probability $\frac{exp(policy[code(move)])}{z}$
12:       *state* ← play (*state*, *move*)
13:       *sequence* ← *sequence* + *move*
14:   **end while**

The problem is even worse when this situation occurs at the beginning of a nested level since there is not yet a best sequence. In this case the policy is adapted directly on this bad sequence, pushing the rollouts towards bad sequences, which perturbs the rollouts of the entire nested level.

To prevent this problem, an idea is simply to generate not only 1 sequence according to a given policy, but P sequences, in order to get a better evaluation

Algorithm 2. The Adapt algorithm

1: Adapt (*policy, sequence*)
2:    *polp* ← *policy*
3:    *state* ← *root*
4:    **for** *move* in *sequence* **do**
5:        *polp* [code(*move*)] ← *polp* [code(*move*)] + α
6:        z ← 0.0
7:        **for** m in possible moves for *state* **do**
8:            z ← z + exp (*policy* [code(*m*)])
9:        **end for**
10:       **for** m in possible moves for *state* **do**
11:           *polp* [code(*m*)] ← *polp* [code(*m*)] - $\alpha * \frac{exp(policy[code(m)])}{z}$
12:       **end for**
13:       *state* ← play (*state, move*)
14:   **end for**
15:   *policy* ← *polp*
16:   **return** *policy*

Algorithm 3. The NRPA algorithm.

1: NRPA (*level, policy*)
2:    **if** level == 0 **then**
3:        **return** playout (root, *policy*)
4:    **else**
5:        *bestScore* ← $-\infty$
6:        **for** N iterations **do**
7:            (result,new) ← NRPA(*level* − 1, *policy*)
8:            **if** result ≥ bestScore **then**
9:                bestScore ← result
10:               seq ← new
11:           **end if**
12:           policy ← Adapt (policy, seq)
13:       **end for**
14:       **return** (bestScore, seq)
15:   **end if**

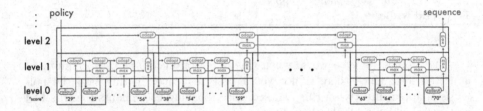

Fig. 1. NRPA scheme

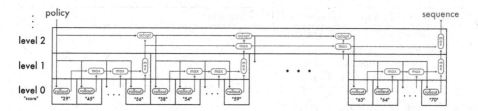

Fig. 2. Stabilized NRPA scheme. Level 1 is replaced by an evaluation level

of the performance of this policy. The algorithm does not adapt to the best sequence until P sequence have been played. And the best sequence returned is the best sequence over those P sequences.

We note that doing so stabilizes the convergence of NRPA. Rollouts are less often pushed to bad sequences, making entire nested level less perturbed, and so making each nested level useful for the search efficiency, leading also to faster convergence.

In our experiments, we have replaced classic level 1 by an evaluation level leading to Fig. 2, that aims to better evaluate the policy, and to return the best sequence found by this policy. We can see in Fig. 2 that multiple level zero calls are performed before doing the adapt in green whereas in Fig. 1 the green adapt function is called after every level zero call.

The number of evaluation is parameterized by the P parameter and the number of playouts at the lowest level of SNRPA is P times greater than the number of playout at the lowest level of NRPA.

Note that for a fixed number of playouts, the Stabilized NRPA makes less updates comparing to NRPA, making it faster. Note further that Stabilized NRPA is a generalization of NRPA, since SNRPA(1) is NRPA.

Stabilized NRPA is given in Algorithm 4. It follows the same pattern as NRPA. Lines 2–3 and lines 14–25 are the same as in NRPA. They correspond to level zero and to levels strictly greater than one. The difference lies in level one (lines 4–13). At level one there is an additional loop from 1 to P that gets the best sequence out of P playouts.

3.2   Parallelization

Parallelizing NMCS was done in [9]. Parallelizing NRPA on a cluster is easily done using root parallelization when distributing among the different computers and using leaf parallelization on each multiple cores computer [22]. More recently Andrzej Nagorko efficiently parallelized NRPA while not changing its global behavior [21].

Stabilized NRPA is well fitted for leaf parallelization as the P playouts can be done in parallel.

Algorithm 4. The Stabilized NRPA algorithm.

1: StabilizedNRPA (*level*, *policy*)
2:    **if** level == 0 **then**
3:        **return** playout (root, *policy*)
4:    **else if** level == 1 **then**
5:        *bestScore* ← −∞
6:        **for** 1, ..., P **do**
7:            (result,new) ← StabilizedNRPA(*level* − 1, *policy*)
8:            **if** result ≥ bestScore **then**
9:                bestScore ← result
10:               seq ← new
11:            **end if**
12:        **end for**
13:        **return** (bestScore, seq)
14:    **else**
15:        *bestScore* ← −∞
16:        **for** 1, ..., N **do**
17:            (result,new) ← StabilizedNRPA(*level* − 1, *policy*)
18:            **if** result ≥ bestScore **then**
19:                bestScore ← result
20:               seq ← new
21:            **end if**
22:            policy ← Adapt (policy, seq)
23:        **end for**
24:        **return** (bestScore, seq)
25:    **end if**

---

4   Problems Used for the Experiments

In this section we present the three problems used for the experiments. The Maximum problem where the goal is to find a mathematical expression that evaluates as great as possible. The TSPTW problem that finds short paths to visit as set of cities with time constraints. The SameGame problem, a popular puzzle.

4.1   The Maximum Problem

Nested Monte Carlo Search can be used for the optimization of mathematical expressions [5,6,8]. For some problems it gives better results than alternative algorithms such as UCT [17] or Genetic Programming [18].

The Maximum problem [19] consists in finding an expression that results in the maximum possible number given some limit on the size of the expression. In the experiment limit was on the depth of the corresponding tree and the available atoms were +, * and 0.5. In our experiments we fixed a limit on the number of atoms of the generated expression, not on the depth of the tree and the available atoms are +, * and 1.0.

We applied NRPA to the Maximum Problem. It is the first time NRPA is applied to Expression Discovery.

Figure 3 gives an example of how an expression is built using a playout. The left tree corresponds to the stack of atoms below the tree. The stack defines a tree and in order to fill the tree new atoms are pushed on top of the stack. For example pushing the '+' atom on the stack gives the tree on the right. When the maximum number of nodes + leaves is reached for a stack only terminal atoms (atoms that do not have children) are pushed onto the stack enforcing the number of nodes of the generated expression to be below the defined maximum.

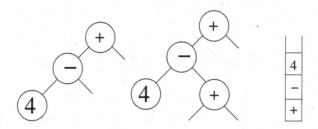

Fig. 3. A partial tree and the corresponding stack.

4.2 TSPTW

In the Traveling Salesman Problem with Time Windows (TSPTW) an agent has to visit N cities at predefined periods of times while minimizing the total tour cost. NRPA has been successfully applied to TSPTW [11,12].

The Hamiltonian Path problem is a subproblem of the TSP, so TSPTW and most other TSP variants are computationally hard. No algorithm polynomial in the number of cities is expected.

The TSPTW is much harder than the TSP, different algorithms have to be used for solving this problem and NRPA had state of the art results on standard benchmarks.

Following the formulation of [11], the TSPTW can be defined as follow. Let G be an undirected complete graph. $G = (N, A)$, where $N = 0, 1, \ldots, n$ corresponds to a set of nodes and $A = N \times N$ corresponds to the set of edges between the nodes. The node 0 corresponds to the depot. Each city is represented by the n other nodes. A cost function $c : A \rightarrow R$ is given and represents the distance between two cities. A solution to this problem is a sequence of nodes $P = (p_0, p_1, \ldots, p_n)$ where $p_0 = 0$ and $(p_1, \ldots, p_n)$ is a permutation of $[1, N]$. Set $p_{n+1} = 0$ (the path must finish at the depot), then the goal is to minimize the function defined in Eq. 1.

$$cost(P) = \sum_{k=0}^{n} c(a_{p_k}, a_{p_{k+1}}) \tag{1}$$

As said previously, the TSPTW version is more difficult because each city i has to be visited in a time interval $[e_i, l_i]$. This means that a city i has to be visited before l_i. It is possible to visit a cite before e_i, but in that case, the new departure time becomes e_i. Consequently, this case may be dangerous as it generates a penalty. Formally, if r_{p_k} is the real arrival time at node p_k, then the departure time d_{p_k} from this node is $d_{p_k} = max(r_{p_k}, e_{p_k})$.

In the TSPTW, the function to minimize is the same as for the TSP (Eq. 1), but a set of constraint is added and must be satisfied. Let us define $\Omega(P)$ as the number of violated windows constraints by tour (P).

Two constraints are defined. The first constraint is to check that the arrival time is lower than the fixed time. Formally,

$$\forall p_k, r_{p_k} < l_{p_k}.$$

The second constraint is the minimization of the time lost by waiting at a city. Formally,

$$r_{p_{k+1}} = max(r_{p_k}, e_{p_k}) + c(a_{p_k, p_{k+1}}).$$

In NRPA paths with violated constraints can be generated. As presented in [27] , a new score $Tcost(p)$ of a path p can be defined as follow:

$$Tcost(p) = cost(p) + 10^6 * \Omega(p),$$

with, as defined previously, $cost(p)$ the cost of the path p and $\Omega(p)$ the number of violated constraints. 10^6 is a constant chosen high enough so that the algorithm first optimizes the constraints.

The problem we use to experiment with the TSPTW problem is the most difficult problem from the set of [24].

4.3   SameGame

In SameGame the goal is to score as much as possible removing connected components of the same color. An example of a SameGame board is given in Fig. 4. The score for removing n tiles is $(n - 2)^2$. If the board is completely cleared there is a bonus of 1000.

When applying Monte Carlo Search to SameGame it is beneficial to use selective search [7] in order to eliminate moves that are often bad. For example it is important to remove the tiles of the dominant color all at once in order to score a lot with this move. The Tabu color playout strategy achieves this by forbidding moves of the dominant color when they do not clear all the tiles of the dominant color in one move. We sometimes allow moves of size two for the dominant color beside the Tabu color strategy as advised in [7].

The best published results for SameGame come from a parallel implementation of NRPA [22].

Figure 4 gives the first problem of the standard SameGame suite. This is the one we used in our experiments.

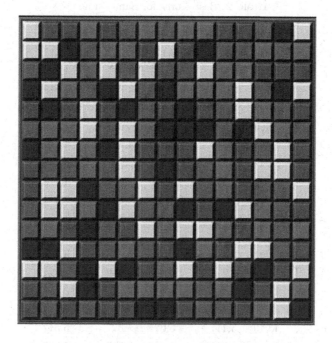

Fig. 4. First problem of the SameGame standard suite

Table 1. Results for the Maximum problem (scale × 1 000 000).

Time	NRPA	SNRPA(2)	SNRPA(3)	SNRPA(4)	SNRPA(10)
0.01	1	1	**2**	**2**	1
0.02	3	24	68	138	25
0.04	16	108	326	934	**5086**
0.08	150	1341	2092	5971	**21745**
0.16	3475	15844	19874	52041	**88547**
0.32	170265	534672	487983	**1147083**	789547
0.64	13803062	28885199	22863271	**36529536**	12000748
1.28	40077774	216376610	270326701	379573875	212668695
2.56	89668935	314740908	408327339	495249021	**708820733**
5.12	151647343	472960557	557957691	704240083	**904642720**
10.24	345707890	712902227	856149587	938008979	**1296603114**
20.48	852761999	1151948749	1284225823	1359946097	**1661398711**
40.96	1975250628	2168737831	2221426342	**2232301333**	2128244879
81.92	2973605038	3276850130	**3381032884**	3321287204	3057041220
163.84	3336604131	3531572024	3627351674	3621195107	**3928494648**

Table 2. Best scores for SameGame

Problem	NMCS	SP-MCTS	NRPA	SRNPA(4)	js-games.de
1	3,121	2,919	3,179	3,203	3,413
2	3,813	3,797	3,985	3,987	4,023
3	3,085	3,243	3,635	3,757	4,019
4	3,697	3,687	3,913	4,001	4,215
5	4,055	4,067	4,309	4,287	4,379
6	4,459	4,269	4,809	4,799	4,869
7	2,949	2,949	2,651	2,151	3,435
8	3,999	4,043	3,879	4,079	4,771
9	4,695	4,769	4,807	4,821	5,041
10	3,223	3,245	2,831	3,333	3,937
11	3,147	3,259	3,317	3,531	3,783
12	3,201	3,245	3,315	3,355	3,921
13	3,197	3,211	3,399	3,379	3,821
14	2,799	2,937	3,097	3,121	3,263
15	3,677	3,343	3,559	3,783	4,161
16	4,979	5,117	5,025	5,377	5,517
17	4,919	4,959	5,043	5,049	5,227
18	5,201	5,151	5,407	5,491	5,503
19	4,883	4,803	5,065	5,325	5,343
20	4,835	4,999	4,805	5,203	5,217
Total	77,934	78,012	80,030	82,032	87,858

5   Experimental Results

In all our experiments we use the average score over 200 runs. For each search time starting at 0.01 s and doubling until 163.84 s we give the average score reached by Stabilized NRPA within this time. We run a search of level 4 each run, but NRPA does not has the time to finish level 3, especially when running SNRPA(P). SNRPA(P) advances approximately P times less steps than NRPA at level 3 since it spends approximately P times more at level 1. All the experiments use sequential versions of NRPA and Stabilized NRPA.

Table 1 gives the evolution for the Maximum problem. The score is the evaluation of the mathematical expression. The first column gives the average scores of standard NRPA. The second column gives the average scores of Stabilized NRPA with $P = 2$. The third column gives the average scores of Stabilized NRPA with $P = 3$ and so on. We can observe that SNRPA(10) gives the best results. To save place the numbers generated by the expressions have been divided by 1 000 000.

Table 3. Results for the TSPTW rc204.1 problem

Time	NRPA	SNRPA(2)	SNRPA(3)	SNRPA(4)	SNRPA(10)
0.01	−29037026	**−28762022**	−29107010	−29222032	−29337060
0.02	−26501832	−26121858	−26226870	−26181904	−27096936
0.04	−25276756	−24221694	−24056722	−23596696	−24031802
0.08	−23821720	−22621656	−22556632	−22176624	−21706624
0.16	−22006640	−21436606	−21216568	−20806566	−20261500
0.32	−19521526	−19441520	−19481502	−19086484	−18821438
0.64	**−16416390**	−16536396	−16536403	−16536387	−17166394
1.28	−13966259	−13636262	−13466266	**−13316265**	−14691306
2.56	−12781221	−11881189	−11111173	−10856164	−11696195
5.12	−11301179	−10556154	−9866131	−9406120	**−8831112**
10.24	−9351129	−8816107	−8166091	−7866081	**−7241065**
20.48	−6591049	−6631047	−6166038	**−6031033**	−6076040
40.96	**−3695987**	−3890987	−3975989	−4045989	−4085994
81.92	−1825960	−1560955	**−1505955**	−1540954	−2100962
163.84	−980946	−780941	−580938	−500938	**−385937**

Table 4. Results for the first problem of SameGame

Time	NRPA	SNRPA(2)	SNRPA(3)	SNRPA(4)	SNRPA(5)	SNRPA(6)	SNRPA(7)	SNRPA(10)
0.01	448	499	483	494	**504**	479	485	464
0.02	654	685	678	**701**	676	672	660	637
0.04	809	871	863	896	**904**	867	836	823
0.08	927	989	1010	**1062**	**1062**	1045	1032	1026
0.16	1044	1091	1133	1183	1172	1156	1170	**1186**
0.32	1177	1214	1239	1286	1286	1278	1285	**1288**
0.64	1338	1370	1386	1407	**1414**	1396	1403	1398
1.28	1514	1548	1559	1556	**1573**	1544	1548	1547
2.56	1662	1739	1721	1740	**1759**	1713	1733	1716
5.12	1790	1859	1894	1900	1913	**1917**	1913	1897
10.24	1928	2046	2025	2034	2068	**2080**	2071	2065
20.48	2113	2228	2249	**2277**	2255	2271	2243	2213
40.96	2393	2518	2475	**2556**	2518	2513	2471	2477
81.92	2642	2753	2718	**2787**	2761	2760	2733	2700
163.84	2838	2898	2868	**2949**	2940	2945	2912	2917

Table 3 gives the results for the rc204.1 TSPTW problem. This is the most difficult problem of the Solomon-Potwin-Bengio TSPTW benchmark. The score is one million times the number of violated constraints plus the tour cost. SNRPA(10) gives again the best results.

Table 4 gives the results for the first problem of SameGame. Evaluation improves the performance until a certain limit. Indeed, $P = 4$ provides the best results with $P = 5$ and $P = 6$ yielding close scores.

For the three problems, Stabilized NRPA gives better results than NRPA.

Among the different version of SNRPA, the conclusion differs depending of the problem we consider :

For the Maximum Problem, we note that values as great as 10 for P give the best results. For the TSPTW Problem, we note that for the longest time (163.84 s), we go from -980946 for NRPA, to -385937 for SNRPA(10) the best result for the greatest value we have tried for P. On the contrary smaller values for P seem appropriate for SameGame with $P = 4$ being the best.

Table 2 gives the scores reached by different algorithms on the standard test set of 20 SameGame problems. We see that SNRPA(4) improves on NRPA at level 4. However SNRPA(4) takes more time when run sequentially since it does four times more playouts as NRPA. Still is does the same number of calls to the adapt function as NRPA. SP-MCTS is a variant of the UCT algorithm applied to single player games, and NMCS is Nested Monte Carlo Search. They both reach smaller overall scores than SNRPA(4). The last column contains the records from the website js-games.de. They were obtained by specialized algorithms and little is known about these algorithms except that some of them use a kind of beam search with specialized evaluation functions.

6   Conclusion

Stabilized NRPA is a simple modification of the NRPA algorithm. It consists in periodically playing P playouts at the lowest level before performing the adaptation. It is a generalization of NRPA since Stabilized NRPA with $P = 1$ is NRPA. It improves the average scores of NRPA given the same computation time for three different problems: Expression Discovery, TSPTW and SameGame.

Acknowledgment. This work was supported in part by the French government under management of Agence Nationale de la Recherche as part of the "Investissements d'avenir" program, reference ANR19-P3IA-0001 (PRAIRIE 3IA Institute).

References

1. Bouzy, B.: Monte-Carlo fork search for cooperative path-finding. In: Computer Games - Workshop on Computer Games, CGW 2013, Held in Conjunction with the 23rd International Conference on Artificial Intelligence, IJCAI 2013, Beijing, China, 3 August 2013, Revised Selected Papers, pp. 1–15 (2013)
2. Bouzy, B.: Burnt pancake problem: new lower bounds on the diameter and new experimental optimality ratios. In: Proceedings of the Ninth Annual Symposium on Combinatorial Search, SOCS 2016, Tarrytown, NY, USA, 6–8 July 2016, pp. 119–120 (2016)
3. Browne, C., et al.: A survey of Monte Carlo tree search methods. IEEE Trans. Comput. Intell. AI Games 4(1), 1–43 (2012). https://doi.org/10.1109/TCIAIG. 2012.2186810
4. Cazenave, T.: Nested Monte-Carlo search. In: Boutilier, C. (ed.) IJCAI, pp. 456–461 (2009)

5. Cazenave, T.: Nested Monte-carlo expression discovery. In: ECAI 2010–19th European Conference on Artificial Intelligence, Lisbon, Portugal, 16–20 August 2010, Proceedings, pp. 1057–1058 (2010). https://doi.org/10.3233/978-1-60750-606-5-1057

6. Cazenave, T.: Monte-carlo expression discovery. Int. J. Artif. Intell. Tools **22**(1) (2013). https://doi.org/10.1142/S0218213012500352

7. Cazenave, T.: Nested rollout policy adaptation with selective policies. In: CGW at IJCAI 2016 (2016)

8. Cazenave, T., Hamida, S.B.: Forecasting financial volatility using nested Monte Carlo expression discovery. In: IEEE Symposium Series on Computational Intelligence, SSCI 2015, Cape Town, South Africa, 7–10 December 2015, pp. 726–733 (2015). https://doi.org/10.1109/SSCI.2015.110

9. Cazenave, T., Jouandeau, N.: Parallel nested monte-carlo search. In: 23rd IEEE International Symposium on Parallel and Distributed Processing, IPDPS 2009, Rome, Italy, 23–29 May 2009, pp. 1–6 (2009). https://doi.org/10.1109/IPDPS.2009.5161122

10. Cazenave, T., Saffidine, A., Schofield, M.J., Thielscher, M.: Nested Monte Carlo search for two-player games. In: Proceedings of the Thirtieth AAAI Conference on Artificial Intelligence, 12–17 February 2016, Phoenix, Arizona, USA, pp. 687–693 (2016). http://www.aaai.org/ocs/index.php/AAAI/AAAI16/paper/view/12134

11. Cazenave, T., Teytaud, F.: Application of the nested rollout policy adaptation algorithm to the traveling salesman problem with time windows. In: Hamadi, Y., Schoenauer, M. (eds.) LION 2012. LNCS, pp. 42–54. Springer, Heidelberg (2012). https://doi.org/10.1007/978-3-642-34413-8_4

12. Edelkamp, S., Gath, M., Cazenave, T., Teytaud, F.: Algorithm and knowledge engineering for the TSPTW problem. In: 2013 IEEE Symposium on Computational Intelligence in Scheduling (SCIS), pp. 44–51. IEEE (2013)

13. Edelkamp, S., Gath, M., Greulich, C., Humann, M., Herzog, O., Lawo, M.: Monte-Carlo tree search for logistics. In: Clausen, U., Friedrich, H., Thaller, C., Geiger, C. (eds.) Commercial Transport. LNL, pp. 427–440. Springer, Cham (2016). https://doi.org/10.1007/978-3-319-21266-1_28

14. Edelkamp, S., Gath, M., Rohde, M.: Monte-Carlo tree search for 3D packing with object orientation. In: Lutz, C., Thielscher, M. (eds.) KI 2014. LNCS (LNAI), vol. 8736, pp. 285–296. Springer, Cham (2014). https://doi.org/10.1007/978-3-319-11206-0_28

15. Edelkamp, S., Greulich, C.: Solving physical traveling salesman problems with policy adaptation. In: 2014 IEEE Conference on Computational Intelligence and Games (CIG), pp. 1–8. IEEE (2014)

16. Edelkamp, S., Tang, Z.: Monte-Carlo tree search for the multiple sequence alignment problem. In: Eighth Annual Symposium on Combinatorial Search (2015)

17. Kocsis, L., Szepesvári, C.: Bandit based Monte-Carlo planning. In: Fürnkranz, J., Scheffer, T., Spiliopoulou, M. (eds.) ECML 2006. LNCS (LNAI), vol. 4212, pp. 282–293. Springer, Heidelberg (2006). https://doi.org/10.1007/11871842_29

18. Koza, J.R., et al.: Genetic programming II, vol. 17. MIT Press, Cambridge (1994)

19. Langdon, W.B., Poli, R., et al.: An analysis of the max problem in genetic programming. Genet. Program. **1**(997), 222–230 (1997)

20. Méhat, J., Cazenave, T.: Combining UCT and nested Monte Carlo search for single-player general game playing. IEEE Trans. Comput. Intell. AI Games **2**(4), 271–277 (2010)

21. Nagorko, A.: Parallel nested rollout policy adaptation. In: IEEE Conference on Games (2019)

22. Négrevergne, B., Cazenave, T.: Distributed nested rollout policy for samegame. In: Computer Games - 6th Workshop, CGW 2017, Held in Conjunction with the 26th International Conference on Artificial Intelligence, IJCAI 2017, Melbourne, VIC, Australia, 20 August 2017, Revised Selected Papers, pp. 108–120 (2017). https://doi.org/10.1007/978-3-319-75931-9_8

23. Portela, F.: An unexpectedly effective Monte Carlo technique for the RNA inverse folding problem. bioRxiv, p. 345587 (2018)

24. Potvin, J.Y., Bengio, S.: The vehicle routing problem with time windows part II: genetic search. INFORMS J. Comput. 8(2), 165–172 (1996)

25. Poulding, S.M., Feldt, R.: Generating structured test data with specific properties using nested Monte-Carlo search. In: Genetic and Evolutionary Computation Conference, GECCO 2014, Vancouver, BC, Canada, 12–16 July 2014, pp. 1279–1286 (2014)

26. Poulding, S.M., Feldt, R.: Heuristic model checking using a Monte-Carlo tree search algorithm. In: Proceedings of the Genetic and Evolutionary Computation Conference, GECCO 2015, Madrid, Spain, 11–15 July 2015, pp. 1359–1366 (2015)

27. Rimmel, A., Teytaud, F., Cazenave, T.: Optimization of the nested Monte-Carlo algorithm on the traveling salesman problem with time windows. In: Di Chio, C., et al. (eds.) EvoApplications 2011, Part II. LNCS, vol. 6625, pp. 501–510. Springer, Heidelberg (2011). https://doi.org/10.1007/978-3-642-20520-0_51

28. Rosin, C.D.: Nested rollout policy adaptation for Monte Carlo tree search. In: IJCAI, pp. 649–654 (2011)

zoNNscan: A Boundary-Entropy Index for Zone Inspection of Neural Models

Adel Jaouen[1] and Erwan Le Merrer[2(✉)]

[1] Technicolor, Paris, France
[2] Univ Rennes, Inria, CNRS, Irisa, Rennes, France

Abstract. The training of deep neural network classifiers results in decision boundaries whose geometry is still not well understood. This is in direct relation with classification problems such as so called *corner case* inputs. We introduce zoNNscan, an index that is intended to inform on the boundary uncertainty (in terms of the presence of other classes) around one given input datapoint. It is based on confidence entropy, and is implemented through Monte Carlo sampling in the multidimensional ball surrounding that input. We detail the zoNNscan index, give an algorithm for approximating it, and finally illustrate its benefits on three applications.

Keywords: Decision boundaries · Classification uncertainty · Entropy · Application of Monte Carlo sampling

1 Introduction

Measures provide crucial insights in all computer science domains, as far as they allow to quantify and then analyze specific problems [18], or to serve as a building block for designing better algorithms [8]. In the blooming field of machine learning leveraging deep neural networks, open problems remain that are related to the position of *decision boundaries* in classification tasks. Little is known about the geometrical properties of those [4,6,16]. Yet, the close proximity of some inputs to those boundaries is getting more problematic with the increasing amount of critical applications that are using deep neural networks. The case of self-driving cars is an illustration of a critical application, where corner-cases have been recently found in production models (leading to wrong decisions) [17]; the inputs that cause the erroneous classifications are depicted to be close to decision boundaries. There are also recent works that are voluntarily tweaking the decision boundaries around given inputs, so that ownership information can be embedded into a target model [11].

While measures are inherently in use with neural networks to evaluate the quality of the learned model over a given dataset, we find that there is a lack of an index that provides information on the neighborhood of given inputs, with regards to the boundaries of other classes. Visual inspections of suspicious zones, by plotting in 2D a given decision boundary and the considered inputs is possible

T. Cazenave et al. (Eds.): MCS 2020, CCIS 1379, pp. 31–38, 2021.
https://doi.org/10.1007/978-3-030-89453-5_3

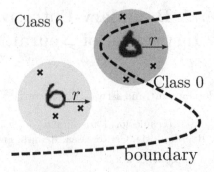

Fig. 1. An illustration of zoNNscan runs around two inputs. The example on the left (a 6 digit) is far from the model decision boundary, thus a zoNNscan at radius r returns a value close to 0. The example on the right is another 6 from the MNIST dataset, that is often misclassified (here as a 0, such as in [3]); its is close to the decision boundary of the 6 class. zoNNscan returns a higher value (up to 1) for it, as some data samples in the ball fall into the correct digit class.

[2]. In this paper, we propose an algorithm to output an index, for it helps to quantify possible problems at inference time.

2   Zoning Classification Uncertainty with zoNNscan

Classically, given an input and a trained model, the class predicted by the classifier is set to be the one with the higher score in the output vector [4]. These scores can thus be considered as membership probabilities to the model classes: output values close to 1 are to be interpreted as a high confidence about the prediction, and conversely, values close to 0 show high probabilities of non membership to the corresponding class. These scores also provide the information about the *uncertainty* of the model for a given input : vector scores close to each other indicate an uncertainty between these classes, and equal scores characterize a decision boundary point [13]. Thereby, a maximum uncertainty output refers to an input that causes inference to return an uniform vector of probabilities $\frac{1}{C}$ for a classification task into C-classes. Conversely, minimum uncertainty corresponds to an input that causes inference to return a vector of zeros, except for the predicted class for which a one is set. zoNNscan evaluates the uncertainty of predictions in a given input region. This is illustrated on Fig. 1.

2.1   zoNNscan: Definition

Given a discrete probability distribution $p = \{p_1, ..., p_C\} \in [0, 1]^C$ with $\Sigma_{i=1}^C p_i = 1$ (C designates the number of events), we remind the definition of the Shannon entropy (b designates the base for the logarithm) :

$$H_b : [0, 1]^C \rightarrow \mathbb{R}^+$$
$$p \quad \mapsto \Sigma_{i=1}^C (-p_i \log_b(p_i)).$$

The maximum of H_b is reached in $p = \{\frac{1}{C}, ..., \frac{1}{C}\}$ and is equal to $\log_b(C)$ and with the convention $0\log_b(0) = 0$, the minimum of H_b is 0. Subsequently, we use a C-based logarithm to make H_C output values in range $[0, 1]$. A classifier $\mathcal{M}$ is considered as a function defined on the input space $[0, 1]^d$ (on the premise that data are generally normalized for the training phase in such a way that all features are included in the range $[0, 1]$), taking its values in the space $\mathbb{P} = \{p \in [0, 1]^C \text{ s.t. } \Sigma_{i=1}^C p_i = 1\}$. We introduce the composite function $\varphi = H_b \circ \mathcal{M}$ to model the indecision the network has on the input space. More specifically, we propose the expectation of $\varphi(U)$, $\mathbb{E}_{\mathbb{Z}}[\varphi(U)]$ (with U a uniform random variable on $\mathbb{Z}$), on an input zone $\mathbb{Z} \in [0, 1]^d$, to be an indicator on the uncertainty of the classifier on zone $\mathbb{Z}$.

Definition 1. *Let the zoNNscan index be, in zone $\mathbb{Z}$:*

$$\mathcal{B}([0, 1]^d) \rightarrow [0, 1]$$
$$\mathbb{Z} \qquad \mapsto \mathbb{E}_{\mathbb{Z}}[\varphi(U)] = \int_{\mathbb{Z}} \varphi(u) f_U(u)\, du,$$

where $\mathcal{B}(\mathbb{R}^d)$ refers to the $\mathbb{R}^d$ Borel set and f_U the uniform density on $\mathbb{Z}$.

The minima of $\mathbb{E}_{\mathbb{Z}}[\varphi(U)]$ depicts observations in $\mathbb{Z}$ that were all returning one confidence of 1, and $C - 1$ confidences of 0. Conversely, the maxima indicates full uncertainty, where each observation returned $\frac{1}{C}$ confidence values in the output vector.

2.2 A Monte Carlo Approximation of zoNNscan

In practice, as data are typically nowadays of high dimensionality (e.g., in the order of millions of dimensions for image applications) and deep neural networks are computing complex non-linear functions, one cannot expect to compute the exact value of this expectation. We propose a Monte Carlo method to estimate this expectation on a space corresponding to the surrounding zone of a certain input.

For inspection around a given input $X \in [0, 1]^d$, we consider a ball **B** of center X and radius r, as zone $\mathbb{Z}$ introduced in previous subsection. We perform a Monte Carlo sampling of k inputs in a ball for the infinite-norm, corresponding to the hyper-cube of dimension d, around X (as depicted on Fig. 1). We are generating inputs applying random deformations[1] ϵ_i on each components X_i such as $max(-X_i, -r) \le \epsilon_i \le min(1 - X_i, r)$.

For instance, given a normalized input X and a positive radius r, Monte Carlo sampling is performed uniformly in the subspace of $\mathbb{R}^d$ defined as:

$$\mathbb{Z} = \mathbf{B}_\infty(X, r) \cap [0, 1]^d.$$

---

[1] For more advanced sampling techniques in high dimensional spaces, please refer to e.g., [5].

Algorithm 1. Monte Carlo approximation of `zoNNscan`

Require: $\mathcal{M}, X, r, k$
  for $i = 0..k$ **do**
    $x' \leftarrow$ Monte Carlo Sampling in $\mathbf{B}_\infty(X, r) \cap [0, 1]^d$
    $P_i \leftarrow \mathcal{M}(x')$ {Inference}
    $entropy_i \leftarrow H_C(P_i)$
  end for
  return $\frac{1}{k} \sum_{i=1}^{k} entropy_i$

The larger the number of samples k, the better the approximation of the index; this value has to be set considering the application and the acceptable computation/time constraints for inspection.

The `zoNNscan` index, for inspection of the surrounding of input X in a model $\mathcal{M}$, is presented on Algorithm 1.

3 Illustration Use of `zoNNscan`

We now look at three example use cases for `zoNNscan`.

3.1 Uncertainty Around a Given Input

A first `zoNNscan` use case is to assess at which distance and in which proportions are other classes surrounding an interesting input X.

We experiment using the multi-layer perceptron (noted MLP hereafter) proposed as an example in the Keras library [1], and with $k = 1,000$ for the remaining of the experiments. Figure 2 presents the `zoNNscan` values with respect various radii r, around two input examples (a 6 digit from MNIST and the image of a ship from the CIFAR-10 dataset). A radius value $r = 0$ is equivalent to the simple output vector corresponding to the inference of X, i.e., without actual sampling in $\mathbf{B}$. Our algorithm starts with a $r = 0$, up to 1^2. For the top-experiment on the MNIST digit, the confidence value for $r = 0$ is high for digit 6 (around 0.9), which is reflected by `zoNNscan` value to be low (around 0.2). When r increases, the confidence values of other classes are increasing progressively, which results for `zoNNscan` increasing sharply; at radii of $r = [0.25, 1]$ uncertainty in such a ball is critical.

For the bottom-experiment on the ship image the CIFAR-10 dataset, we observe at $r = 0$ and close a higher value (around 0.25), that captures the relatively high score for the class 'cat'. The score remains around 0.25 up to $r = 0.30$, because other boundaries ('frog' and 'truck') are nearby. Finally `zoNNscan` tends to zero because the 'truck' class occupies the input space in its vast majority. This last fact illustrates that `zoNNscan` is based on probability vectors, and thus

---

^{2} Note that for $X \in [0, 1]^d$ and for $r \geq 1$, $\mathbf{B}_\infty(X, r) \cap [0, 1]^d = [0, 1]^d$, then the space of normalized data is totally covered by the sampling process.

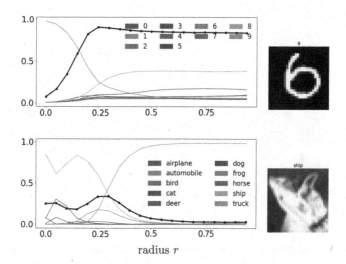

Fig. 2. zoNNscan values (dotted black line) on y-axes, with a varying radius $r \in [0, 1]$ on x-axes. Mean confidence scores for each class in the explored zone are also reported (colored lines). Top: MNIST dataset; a 6 digit classified by the MLP model. Bottom: CIFAR-10 dataset; a ship classified by RESNETv2. (Color figure online)

follows the dominant class in them. For the use of zoNNscan at inference time, it is of course more desirable to have low r values, so that the inspected zone remains confined around the input to examine.

This experiment brings two observations: (i) the study of the evolution of zoNNscan values with growing radii is of interest to characterize surrounding or problematic zones (e.g., sharp increase of the 'truck' class starting at 0.25 on Fig. 2). (ii) for a r covering the whole input space, we obtain an estimate of the surface occupied by each class in the model (e.g., class 8 dominates on MNIST, while class 'truck' is covering most of classification space for the CIFAR-10 case).

3.2 Disagreement on Corner Case Inputs

Arguably interesting inputs are the ones that make several deep neural network models disagree, as depicted in the case of self-driving cars [17]. Indeed, they constitute corner cases, that have to be dealt with for most critical applications. While this problem may be masked using voting techniques by several independent classifiers [9], we are interested in studying the zoNNscan value of those particular inputs.

We add two models for the experiment, also from Keras [1]: a convolutional neural network (CNN), and a recurrent neural network (IRNN), and focus on the MNIST dataset. Given two models, those inputs are extracted by comparing the predictions of the two models on a given dataset. In the MNIST dataset, we identify a total of 182 disagreements (i.e., corner cases) for the three models.

We plot on Fig. 3 the two following empirical distributions. First, in blue, the distribution of 1,000 zoNNscan values around test set inputs; second, in red, the

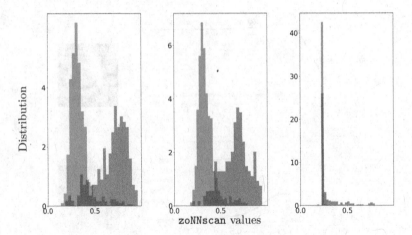

Fig. 3. Relative distributions of zoNNscan around 1,000 random test set inputs (blue) and the 182 corner case inputs (red). From left to right: the MLP, CNN and IRNN models. (Color figure online)

distribution of the 182 corner case inputs. The zoNNscan index is also estimated with $r = 0.025$. We observe a significant difference in the distributions (means are reported on Table 1).

In order to formally assess this difference in the two distributions, we performed the statistical *Kolmogorov-Smirnov 2-sample test* [15]. It tests the null hypothesis that two samples were drawn from the same continuous distribution. The p-values of this test correspond to the probabilities of observing the values under the null hypothesis. In any of the three case (MLP, CNN and IRNN models), the test rejects the null hypothesis with p-values lower than 10^{-3}.

3.3   Effects of Watermarking on Model Boundaries

As a final illustration, we consider the recent work [11] consisting in *watermarking* a model. This operation is performed as follows: first, a watermark key consisting of inputs is created. Half of the key inputs are actually *adversarial examples* [7]. Second, the watermarking step consists in finetuning the model by retraining it over the key inputs, in order to re-integrate the adversarial examples in their original class. That last step is by definition moving the model decision boundaries, which constitutes the watermark that the model owner has sought to introduce.

All three models are watermarked with keys of size 100. Each input of those three keys is used as input X for zoNNscan, for 100 runs of Algorithm 1 with $k = 1,000$ samples each, $r = 0.05$. Two distributions of zoNNscan values are computed: one for the un-marked model, and another for the watermarked one.

If the watermarked model were indistinguishable from the original one, the two distributions would be the same. Once again, the Kolmogorov-Smirnov 2-sample test returns p-values lower than 10^{-3}. This reflects the fact that boundaries have changed around the key inputs, at least enough to be observed by zoNNscan. As the watermarking process is expected to be as stealthy possible, this highlights that the key is to remain a secret for the model owner only [11].

Table 1. Mean of the empirical zoNNscan distributions, for the three neural networks.

Scenario	MLP	CNN	IRNN
1,000 inputs from test set	0.3709	0.3457	0.2403
182 corner case inputs	0.6857	0.6564	0.2989

4 Related Works

While some metrics such as distance-from-boundaries [8] have been developed for support vector machines for instance, the difficulty to reason about the classifier decision boundaries of neural networks have motivated research works since decades. Lee et al. [12] proposed the extraction of features from decision boundaries, followed by [13] that targets the same operation without assuming underlying probability distribution functions of the data. An interesting library for visual inspection of boundaries, from low dimensional embedding is available online [2]. Fawzi et al. propose to analyze some geometric properties of deep neural network classifiers [6], including the curvature of those boundaries. Van den Berg [4] studies the decision region formation in the specific case of feedforward neural networks with sigmoidal nonlinearities.

If one can afford to use an *ensemble* of models, as well as joint *adversarial training* to assess the uncertainty of a model's decisions, the work in [10] provides an interesting approach; this makes it a good candidate for offline analysis of decisions. Cross entropy is one of the most basic measure for the training of neural networks. While it evaluates the divergence of the dataset distribution from the model distribution, it is not applied to inspect the confidence in the inference of specific zones in the input space. This paper addresses the relation of the surrounding of a given input with respect to the presence of decision boundaries of other classes; this relation is captured by the zoNNscan index, possibly at runtime, by leveraging the Shannon entropy.

5 Conclusion

We introduced a novel index, zoNNscan, for inspecting the surrounding of given inputs with regards to the boundary entropy measured in the zones of interest. We have presented a Monte Carlo estimation of that index, and have shown its applicability on a base case, on a technique for model watermarking, and on a concern regarding the adoption of current deep neural networks (corner case inputs). As this index is intended to be generic, we expect other applications to leverage it; future works include the study of the link of zoNNscan values with regards to other identified issues such as *trojaning* attacks on neural networks [14], and on the intensification of the Monte Carlo search.

References

1. Keras: Deep learning for humans. https://github.com/keras-team/keras/tree/master/examples. Accessed 25 Feb 2020
2. Plotting high-dimensional decision boundaries. https://github.com/tmadl/highdimensional-decision-boundary-plot. Accessed 1 July 2018
3. Belongie, S., Malik, J., Puzicha, J.: Shape matching and object recognition using shape contexts. IEEE Trans. Pattern Anal. Mach. Intell. **24** (2002). https://doi.org/10.1109/34.993558
4. van den Berg, E.: Some insights into the geometry and training of neural networks. CoRR arXiv:1605.00329 (2016)
5. Dick, J., Kuo, F.Y., Sloan, I.H.: High-dimensional integration: the quasi-monte carlo way. Acta Numer. **22**, 133–288 (2013). https://doi.org/10.1017/S0962492913000044
6. Fawzi, A., Moosavi-Dezfooli, S., Frossard, P., Soatto, S.: Classification regions of deep neural networks. CoRR arXiv:1705.09552 (2017)
7. Goodfellow, I.J., Shlens, J., Szegedy, C.: Explaining and harnessing adversarial examples. In: ICLR (2015)
8. Guo, G., Zhang, H.J., Li, S.Z.: Distance-from-boundary as a metric for texture image retrieval. In: IEEE International Conference on Acoustics, Speech, and Signal Processing, vol. 3, pp. 1629–1632 (2001). https://doi.org/10.1109/ICASSP.2001.941248
9. Kuncheva, L., Whitaker, C., Shipp, C., Duin, R.: Limits on the majority vote accuracy in classifier fusion. Pattern Anal. Appl. **6**(1), 22–31 (2003)
10. Lakshminarayanan, B., Pritzel, A., Blundell, C.: Simple and scalable predictive uncertainty estimation using deep ensembles. In: Guyon, I., et al. (eds.) Advances in Neural Information Processing Systems 30, pp. 6402–6413. Curran Associates, Inc. (2017). http://papers.nips.cc/paper/7219-simple-and-scalable-predictive-uncertainty-estimation-using-deep-ensembles.pdf
11. Le Merrer, E., Pérez, P., Trédan, G.: Adversarial frontier stitching for remote neural network watermarking. Neural Comput. Appl. **32**(13), 9233–9244 (2019). https://doi.org/10.1007/s00521-019-04434-z
12. Lee, C., Landgrebe, D.A.: Feature extraction based on decision boundaries. IEEE Trans. Pattern Anal. Mach. Intell. **15**(4), 388–400 (1993)
13. Lee, C., Landgrebe, D.A.: Decision boundary feature extraction for neural networks. IEEE Trans. Neural Netw. **8**(1), 75–83 (1997). https://doi.org/10.1109/72.554193
14. Liu, Y., et al.: Trojaning attack on neural networks. In: NDSS (2018)
15. Mann, H.B., Whitney, D.R.: On a test of whether one of two random variables is stochastically larger than the other. Ann. Math. Stat. **18**(1), 50–60 (1947). https://doi.org/10.1214/aoms/1177730491
16. Moosavi-Dezfooli, S., Fawzi, A., Fawzi, O., Frossard, P.: Universal adversarial perturbations. In: CVPR (2017)
17. Pei, K., Cao, Y., Yang, J., Jana, S.: Deepxplore: automated whitebox testing of deep learning systems. In: SOSP (2017)
18. Wang, Z., Bovik, A.C.: A universal image quality index. IEEE Signal Process. Lett. **9**(3), 81–84 (2002). https://doi.org/10.1109/97.995823

Ordinal Monte Carlo Tree Search

Tobias Joppen[1][✉] and Johannes Fürnkranz[2]

[1] Technische Universität Darmstadt, Darmstadt, Germany
tjoppen@ke.tu-darmstadt.de
[2] Johannes Kepler Universität Linz, Linz, Austria
juffi@faw.jku.at

Abstract. In many problem settings, most notably in game playing, an agent receives a possibly delayed reward for its actions. Often, those rewards are handcrafted and not naturally given. Even simple terminal-only rewards, like winning equals 1 and losing equals -1, can not be seen as an unbiased statement, since these values are chosen arbitrarily, and the behavior of the learner may change with different encoding. It is hard to argue about good rewards and the performance of an agent often depends on the design of the reward signal. In particular, in domains where states by nature only have an ordinal ranking and where meaningful distance information between game state values is not available, a numerical reward signal is necessarily biased. In this paper we take a look at Monte Carlo Tree Search (MCTS), a popular algorithm to solve MDPs, highlight a reoccurring problem concerning its use of rewards, and show that an ordinal treatment of the rewards overcomes this problem. Using the General Video Game Playing framework we show dominance of our newly proposed ordinal MCTS algorithm over other MCTS variants, based on the novel Borda-UCB bandit algorithm.

1 Introduction

A Markov decision process (MDP) is a popular problem definition where the target is to select the actions that maximize a long-term reward. Most state-of-the-art algorithms assume numerical rewards. In domains like finance, a real-valued reward is naturally given, but many other domains do not have a natural numerical reward representation. In such cases, numerical values are often handcrafted by experts so that they optimize the performance of their algorithms. This process is not trivial, and it is hard to argue about good rewards. Hence, such handcrafted rewards may easily be erroneous and contain biases. For special cases such as domains with true ordinal rewards, it has been shown that it is impossible to create numerical rewards that are not biased. For example, [22] argue that emotions need to be treated as ordinal information.

In fact, it is often hard or impossible to tell whether domains are real-valued or ordinal by nature. Experts may even design handcrafted numerical reward without thinking about alternatives, since using numerical reward is state of the art and most algorithms need them. In this paper we want to emphasize that numerical rewards do not have to be the ground truth and it may be

© Springer Nature Switzerland AG 2021
T. Cazenave et al. (Eds.): MCS 2020, CCIS 1379, pp. 39–55, 2021.
https://doi.org/10.1007/978-3-030-89453-5_4

worth-while for the machine learning community to have a closer look at other options, ordinal being only one of them. Popular examples for ordinal feedback are medical treatment settings with feedback like *patient dead*, *patient cured after one week* and *patient cured after one month*, where you do not want to trade off faster cured patients with dead ones or information retrieval systems with reported relevance values *Irrelevant*, *Partially Relevant* or *Relevant*.

MCTS (Monte Carlo tree search) is a popular algorithm to solve MDPs. MCTS is used in many successful AI systems, such as AlphaGo [17] or top-ranked algorithms in the general video game playing competitions [9,13]. A reoccurring problem of MCTS with limited time resources is its behavior in case of danger: As a running example we look at a generic platform game, where an agent has to jump over deadly gaps to eventually reach the goal at the right. Dying is very bad, and the more the agent proceeds to the right, the better. The problem occurs by comparing the actions *jump* and *stand still* : jumping either leads to a better state than before because the agent proceeded to the right by successfully jumping a gap, or to the worst possible state (*death*) in case the jump attempt failed. Standing still, on the other hand, safely avoids death, but will never advance to a better game state. MCTS averages the obtained rewards gained by experience, which lets it often choose the safer action and therefore not progress in the game, because the (few) experiences ending with its death pull down the average reward of *jump* below the mediocre but steady reward of standing still. Because of this, the behavior of MCTS has also been called *cowardly* in the literature [7,11]. Transferring those platform game experiences into an ordinal scale eliminates the need of meaningful distances.

In this paper, we present an ordinal MCTS algorithm that is able to utilize such ordinal feedback. We call our novel algorithm Ordinal MCTS (O-MCTS) and compare it to different MCTS variants on synthetic data that reflects our motivation as well as using the General Video Game AI (GVGAI) framework [14]. The algorithm is based on a novel ordinal bandit algorithm and leverages it into the area of tree search. Other than existing methods, we completely rely on ordinal information and do not translate those values into a numerical framework.

In the next section we introduce MDPs as our problem definition and PB-MCTS and MCTS as already known solutions to solve those problems. Additionally, we introduce Borda-UCB, a bandit algorithm which is used as the tree policy in our novel algorithm. In Sect. 3, we present our algorithm, followed by experiments to present how our algorithm compares to the existing ones (Sects. 4 and 4.2).

2   Foundations

In this section, we recapitulate the foundations and related work of our algorithm. First, we introduce Markov decision processes (MDP) and its ordinal variation (O-MDP) as our main problem setting, followed from an introduction of Monte Carlo tree search (MCTS) and PB-MCTS, a preference-based MCTS algorithm. Lastly, we introduce Borda-UCB, an ordinal bandit algorithm used in our algorithms tree policy.

2.1   Markov Decision Process

Conventional Monte Carlo tree search assumes a scenario in which an agent moves through a state space by taking different actions. A MDP can be formalized as the following:

- A (finite) set of *states* S.
- A (finite) set of *actions* A the agent can perform. Sometimes, only a subset $A(s) \subset A$ of actions is applicable in a state s.
- A Markovian *state transition* function $\delta(s' \mid s, a)$ denoting the probability that invoking action a in state s leads the agent to state s'.
- A *reward function* $r(s) \in \mathbb{R}$ that defines the reward the agent receives in state s.
- A distribution of *start states* $\mu(s) \in [0, 1]$, defining the probability that the MDP starts in that state. We assume a single start state s_0, with $\mu(s_0) = 1$ and $\mu(s) = 0 \; \forall s \neq s_0$.
- A set of *terminal states* for which $A(s) = \emptyset$. We assume that only terminal states are associated with a non-zero reward.

The task is to learn a *policy* $\pi(a \mid s)$ that defines the probability of selecting an action a in state s. The optimal policy $\pi^*(a \mid s)$ maximizes the expected, cumulative reward

$$
V(s_t) = \mathbb{E}\left[\sum_{t=0}^{\infty} \gamma^t r(s_t)\right],
$$

$$
= r(s_t) + \gamma \int_S \int_A \delta(s_{t+1} \mid a_t, s_t)\pi(a_t \mid s_t)V(s_{t+1}).
$$

(1)

The optimal policy maximizes $V(s_t)$ for all time steps t. Here, $\gamma \in [0, 1)$ is a discount factor, which dampens the influence of later events in the sequence. For finding an optimal policy, one needs to solve the so-called exploration/exploitation problem. The state/action spaces are usually too large to sample exhaustively. Hence, it is required to trade-off the improvement of the current, best policy (exploitation) with an exploration of unknown parts of the state/action space.

We also investigate an ordinal variation of the classical MDP: O-MDP [21] O-MDP uses ordinal rewards and thus does not use numerical definitions of optimality or regret. In this paper, we are interested in the borda winner in case of ordinal rewards. The borda winner maximizes the average chance of beating a uniform randomly chosen competitor arm. This value is called borda score $(B(a)$ - see Sect. 2.4). The regret of playing a non-optimal arm a instead of the borda winner a^* is the borda score difference: $regret_a = B(a^*) - B(a)$. Obviously it is $regret_{a^*} = 0$. Note that it is not possible to use one numerical bandit to optimize the borda score, since it is not the direct reward that is visible for the agent.

Furthermore, the borda score is not only dependent on the distribution of the current arm, but also is dependent on the reward distributions of all other arms (since the borda score is defined on comparisons of those).

2.2 Monte Carlo Tree Search

Monte Carlo tree search (MCTS) is a method for approximating an optimal policy for a MDP. It builds a partial search tree, which is more detailed where the rewards are high. MCTS spends less time evaluating less promising action sequences, but does not avoid them entirely to explore the state space. The algorithm iterates over four steps [5]:

1. *Selection:* Starting from the root node v_0, a *tree policy* traverses to deeper nodes v_k, until a state with unvisited successor states is reached.
2. *Expansion:* One successor state is added to the tree.
3. *Simulation:* Starting from the new state, a so-called *rollout* is performed, i.e., random actions are played until a terminal state is reached or a depth limit is exceeded.
4. *Backpropagation:* The reward of the last state of the simulation is backed up through all selected nodes.

The UCT formula

$$a_v^* = \max_{a \in A_v} \bar{X}_v(a) + 2C\sqrt{\frac{2\ln n_v}{n_v(a)}} \quad . \tag{2}$$

has been derived from the UCB1 algorithm [3] and is used to select the most interesting action a_v^* in a node v by trading off the expected reward estimated as $\bar{X}_v(a) = \sum_{i=0}^{n_v} X_v^{(i)}(a)/n_v(a)$ from $n_v(a)$ samples $X_v^{(i)}(a)$, with an exploration term $\sqrt{2\ln(n_v)/n_v(a)}$. The trade-off parameter C is often set to $C = 1/\sqrt{2}$, which has been shown to ensure convergence for rewards $\in [0,1]$ [12]. In the following, we will often omit the subscript v when it is clear from the context.

2.3 Preference-Based Monte Carlo Tree Search

A version of MCTS that uses preference-based feedback (PB-MCTS) was recently introduced by [10]. In this setting, agents receive rewards in form of preferences over states. Hence, feedback about single states s is not available, a state s can only be compared to another state s', i.e., $s \succ s'$ (dominance), $s' \succ s$, or $s \not\succ s'$ (incomparable).

An iteration of PB-MCTS contains the same abstract steps like MCTS, but their realization differs. First and foremost, it is impossible to use preference information on a vanilla MCTS iteration, since it only samples a single trajectory, whereas a second state is needed for a comparison. Hence, PB-MCTS does not select a single path per iteration but an entire subtree of the search tree. In each of its nodes, two actions are selected that can be compared to each other. For the selection step, a modified version of the dueling bandit algorithm RUCB [23] is used to select actions.

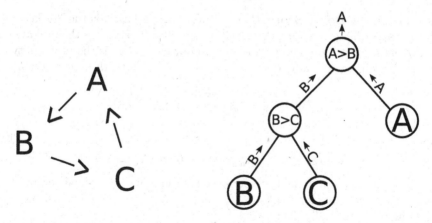

Fig. 1. Three nontransitive actions. The tree introduces a bias to solve nontransitivity.

There are two main disadvantages with this approach:

1. *No transitivity* is used. Given ten actions, MCTS needs only 10 iterations to have a first estimation of quality of each action. In the preference-based approach, each action has to be compared with each other action until a first complete estimation can be done. These are $(10 \cdot 9)/2 = 45$ iterations, i.e., in general the effort is quadratic in the number of actions.
2. A *binary subtree* is needed to learn on each node of the currently best trajectory. Instead of a path of length n for vanilla MCTS, the subtree consists of $2^n - 1$ nodes and 2^{n-1} trajectories instead of only one, causing an exponential blowup of PB-MCTS's search tree.

Hence, we believe that PB-MCTS does not make optimal use of available computing resources, since on a local perspective, transitivity information is lost, and on a global perspective, the desired asymmetric growth of the search tree is undermined by the need for selecting a binary tree. Note that even in the case of a non-transitive domain, PB-MCTS will nevertheless obtain a transitive policy, as illustrated in Fig. 1, where the circular preferences between actions A, B, and C can not be reflected in the resulting tree structure.

2.4 Borda-UCB

A recently proposed algorithm to find the borda winner given a Qualitative Dueling Bandit (QDB) problem is called Borda-UCB. A QDB is a very limited O-MDP with only one non-terminal state, the start state. This problem is not trivial due to stochastic reward distributions of the terminal states.

The base idea of Borda-UCB is to approximate the reward distributions ν_i for each arm a_i. Given those approximate distributions $\hat{\nu}_i$, the pairwise winning probabilities $\mu_{i,j}$ and Borda scores can be approximated.

Estimating the Borda Score B. The Borda score is based on the Borda count which has its origins in voting theory [4]. Essentially, it estimates the probability of winning against a random competitor. In our case, $B(a)$ is the probability of action a to win against any other action $b \neq a$ available (with tie correction):

$$B(a) = \frac{1}{|A| - 1} \sum_{b \in A \setminus \{a\}} \Pr(a \succ b)$$

where

$$\Pr(a \succ b) = \Pr(X_a > X_b) + \tfrac{1}{2} \Pr(X_a = X_b)$$

and X_i is the random variable responsible for sampling the ordinal rewards for arm i. The true value for $\Pr(X_a > X_b)$ and $\Pr(X_a = X_b)$ is unknown but can be estimated empirically [20]:

$$
\begin{aligned}
&\hat{B}(a \succ b) \\
&= \int \Pr[X_a < o]\hat{f}_b(o) + \tfrac{1}{2}\Pr[X_a = o]\hat{f}_b(o)do \\
&= \sum_{i=2}^{|O|} \frac{\hat{F}_a(o_{n-1}) + \hat{F}_a(o_n)}{2} \hat{f}_b(o_n)
\end{aligned}
\tag{3}
$$

Using those pairwise borda score estimates, one can estimate the borda-score $\hat{B}(a)$ of arm a by averaging over all pairwise borda score estimates:

$$\hat{B}(a) = \frac{1}{|A| - 1} \sum_{b \in A \setminus \{a\}} \hat{B}(a \succ b) \tag{4}$$

Borda-UCB Algorithm. In Borda-UCB the Borda Score is used to identify and exploit good performing arms. A common approach to also achieve exploration in bandit algorithms is to use an estimated upper bound of that exploitation term as a selection strategy. The resulting value is high for both, good performing arms as well as for arms with a low sample size while smoothly interpolating between those two extremes[1]:

$$a^* = \arg\max_{a \in A} \hat{B}(a) + 2C\sqrt{\frac{2\ln n}{n(a)}}. \tag{5}$$

where $\alpha > 0$ is a control parameter to trade-off exploration and exploitation. Notice, that the posterior summand (the exploration-term) for action a is not only dependent on the relative number of samples of a, but dependent on the complete sample distribution over all arms. Since the Borda-Score estimate $\hat{B}(a)$ is averaged over multiple pairwise arm comparisons it is necessary to have a good estimate for each pair to receive a tight upper bound for the average.

---

[1] This formula is equivalent to the original Borda-UCB formula after simple algebraic manipulations.

Additionally, the selection strategy of Borda-UCB is slightly different to common state of the art algorithms, too. A common strategy used is to pick the arm that maximizes the upper bound and sample a value for it. Contrary to that, Borda-UCB separates the arms in two sets: The exploitation and exploration set, where the first is composed of all arms having the currently best exploration values and the latter being the other arms. In each time step n we call $a_n = \arg\max Borda\text{-}UCB(a)$ the selected arm, which is played and updated. Instead of moving to the next step, we check whether a_n is an element of the exploration set. If so, the complete exploration set is played sequentially. This way it can be assured that each arm is played regularly and thus the upper bound is not becoming too loose. For a more formal derivation we refer to the original Borda-UCB publication [1].

Our novel O-MCTS algorithm and can be seen as a leverage of Borda-UCB to tree search, just as UCT is a tree search version of UCB1.

3   Ordinal Monte Carlo Tree Search

In this section, we introduce O-MCTS, an MCTS variant which only relies on ordinal information to learn a policy. We derive an ordinal MCTS algorithm by using Borda-UCB instead of UCB as the tree policy.

Ordinal Monte Carlo tree search (O-MCTS) proceeds like conventional MCTS as introduced in Sect. 2.2, but replaces the average value $\bar{X}(a)$ in Eq. (2) with the estimated Borda score $\hat{B}(a)$ used and introduced in the Borda-UCB algorith (Sect. 2.4). Here, each arm is rated according to its mean performance against the other arms. To our knowledge, Borda score has not been used in MCTS before, even though several papers have investigated its use in dueling bandit algorithms [8,16,19]. To calculate the Borda score for each action in a node, O-MCTS stores the backpropagated ordinal values, and estimates pairwise preference probabilities $\hat{B}(a \succ b)$ from these data. Hence, it is not necessary to do multiple rollouts in the same iteration as in PB-MCTS because current rollouts can be directly compared to previously observed ones.

In comparison to UCB, Borda-UCB uses the Borda score $\hat{B}$ as an exploitation term to choose the arm to play (see Formula 5). Note that $\hat{B}(a)$ can only be estimated if each action was visited at least once. Hence, similar to other MCTS variants, we enforce this by always selecting non-visited actions in a node first. Additionally, we did not use the exploitation term or selection strategy used in Borda-UCB, but carryover the common UCT exploration term and selection strategy. We motivate those choices in the following sections.

3.1   Exploration

The Borda-UCB exploration part for action a is composed of two parts: the uncertainty for action a itself and the averaged uncertainty of the other arms (compare Fig. 5). Using $C = 2\frac{|A|-2}{|A|-1}\sqrt{\frac{\alpha}{2}}$ we can show that

$$\sqrt{\frac{\alpha \log n}{n(a)}} + \frac{1}{|A| - 1} \sum_{b \in A \setminus \{a\}} \sqrt{\frac{\alpha \log n}{n(b)}}$$

$$= \frac{|A| - 2}{|A| - 1} \sqrt{\frac{\alpha \log n}{n(a)}} + \frac{1}{|A| - 1} \sqrt{\frac{\alpha \log n}{n(a)}} + \frac{1}{|A| - 1} \sum_{b \in A \setminus \{a\}} \sqrt{\frac{\alpha \log n}{n(b)}}$$

$$= \frac{|A| - 2}{|A| - 1} \sqrt{\frac{\alpha \log n}{n(a)}} + \frac{1}{|A| - 1} \sum_{b \in A} \sqrt{\frac{\alpha \log n}{n(b)}} \qquad (6)$$

$$= \frac{|A| - 2}{|A| - 1} \sqrt{\frac{\alpha \log n}{n(a)}} + R$$

$$= 2C \sqrt{\frac{2 \log n}{n(a)}} + R$$

with $R = \frac{1}{|A| - 1} \sum_{b \in A} \sqrt{\frac{\alpha \log n}{n(b)}}$ being a constant term for all actions a per time step n. Inspecting Formula 6 and considering that we only want to identify the action that maximizes $Borda\text{-}UCB(a)$, we can ignore R since it is the same for all actions. Thus, we can rewrite the exploration part of $Borda\text{-}UCB$ to match Formula 5.

3.2   Tree Policy

As explained in Sect. 2.4, the selection policy of Borda-UCB distinguishes between playing an exploitation action and an exploration action. Whenever an exploration action is chosen, Borda-UCB evaluates all exploration actions at once to ensure a tight upper confidence bound [1]. This can not easily be done in a tree search setting: Evaluating multiple actions per node in a tree can cause an exponential number of paths explored, since this split can happen at each level. The same problem occurred in PB-MCTS and led to performance issues. That is why we chose not to play all exploration actions and only evaluate the one action maximizing Formula 5.

3.3   Discussion

Although the changes from MCTS to O-MCTS are comparably small, the algorithms have very different characteristics. In this section, we highlight some differences between O-MCTS and MCTS.

Different Bias. As mentioned previously, MCTS has been blamed for behaving cowardly, by preferring safe but unyielding actions over actions that have some risk but will in the long run result in higher rewards. As an example, consider Fig. 2, which shows in its bottom row the distribution of trajectory values for two actions over a range of possible rewards. One action (circles) has a mediocre quality with low deviation, whereas the other (stars) is sometimes worse but often

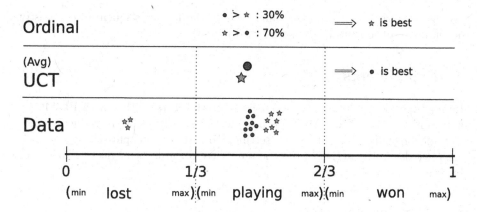

Fig. 2. Two actions with different distributions.

better than the first one. Since MCTS prioritizes the stars only if the average is above the average of circles, MCTS would often choose the safe, mediocre action. In the literature one can find many ideas to tackle this problem, like MixMax backups [11] or adding domain knowledge (e.g., by giving a direct bonus to actions that should be executed [9,13]). O-MCTS takes a different point of view, by not comparing average values but by comparing how often stars are the better option than circles and vice versa. As a result, it would prefer the star action, which is preferable in 70% of the games. Please note that the given example can be inverted such that MCTS takes the right choice instead of O-MCTS.

Hyperparameters and Reward Shaping. When trying to solve a problem with MCTS (and other algorithms, too), rewards can be seen as hyperparameters that can be tuned manually to make an algorithm work as desired. In theory this can be beneficial since you can tweak the algorithm with many parameters. In practice it can be very painful since there often is an overwhelming number of hyperparameters to tune this way. This tuning process is called *reward shaping*. In theory, one can shape the state rewards until a greedy search is able to perform optimal on any problem.

O-MCTS reduces the number of hyperparameters by only asking for ordinal rewards; which is like asking for a ranking of states instead of real numbers for each state. This limits the possibilities of reward shaping but induces a fixed bias using the borda method.

Computational Complexity. Clearly, a naïve computation of $\hat{B}$ is computationally more expensive than MCTS' calculation of a running average. We hence want to point out that once a new ordinal reward is seen it is possible to incrementally update the current value of $\hat{B}$ instead of calculating it again from scratch. In our experiments, updating the Borda score needed 3 to 20 times more time than updating the average (depending on the size of O and A). These values do only show the difference in updating $\hat{B}$ in comparison to updating the running average, not the complete algorithms (where the factor is much lower, mostly depending on the runtime of the forward model).

4 Experiments

In our experiments, we compare our novel O-MCTS algorithm with PB-MCTS and vanilla MCTS. Since MCTS is a numerical algorithm, we introduce numerical and ordinal rewards for each setting. To preserve comparability, the ordinal rewards can be derived from the numerical rewards by neglecting distances between numerical rewards.

4.1 Experimental Setup

We test the three algorithms described above (MCTS, O-MCTS and PB-MCTS) using the General Video Game AI (GVGAI) framework [14]. As additional benchmarks we add MixMax (Q parameter set to 0.25) as an MCTS variation that was suggested by [11] to tackle the cowardly behavior, and Yolobot, a state of the art GVGAI agent [9,13]. GVGAI has implemented a variety of different video games and provides playing agents with a unified interface to simulate moves using a forward model. Using this forward model is expensive so that simulations take a lot of time. We use the number of calls to this forward model as a computational budget. In comparison to using the real computation time, it is independent of specific hardware, algorithm implementations, and side effects such as logging data.

Our algorithms are given access to the following pieces of information provided by the framework:

Available actions: The actions the agent can perform in a given state
Game score: The score of the given state $\in \mathbb{N}$. Depending on the game this
  ranges from 0 to 1 or -1000 to 10000.
Game result: The result of the game: *won, lost* or *running*.
Simulate action: The forward model. It is stochastic, e.g., for enemy moves or
  random object spawns.

Heuristic Monte Carlo Tree Search. The games in GVGAI have a large search space with 5 actions and up to 2000 turns. Using vanilla MCTS, one rollout may use a substantial amount of time, since up to 2000 moves have to be made to reach a terminal state. To achieve a good estimate, many rollouts have to be simulated. Hence it is common to stop rollouts early at non-terminal states, using a heuristic to estimate the value of these states. In our experiments, we use this variation of MCTS, adding the maximal length for rollouts RL as an additional parameter. The heuristic value at non-terminal nodes is computed in the same way as the terminal reward (i.e., it essentially corresponds to the score at this state of the game).

Mapping Rewards to $\mathbb{R}$. The objective function has two dimensions: on the one hand, the agent needs to win the game by achieving a certain goal, on the other hand, the agent also needs to maximize its score. Winning is more important than getting higher scores.

Since MCTS needs its rewards being $\in \mathbb{R}$ or even better $\in [0, 1]$, the two-dimensional target function needs to be mapped to one dimension, in our case for comparison and ease of tuning parameters into $[0, 1]$. Knowing the possible scores of a game, the score can be normalized by $r_{norm} = (r - r_{min})/(r_{max} - r_{min})$ with r_{max} and r_{min} being the highest and lowest possible score.

For modeling the relation $lost \prec playing \prec won$ which must hold for all states, we split the interval $[0, 1]$ into three equal parts (cf. also the axis of Fig. 2):

$$r_{mcts} = \frac{r_{norm}}{3} + \begin{cases} 0, & \text{if } lost \\ \frac{1}{3}, & \text{if } playing \\ \frac{2}{3}, & \text{if } won. \end{cases} \tag{7}$$

This is only one of many possibilities to map the rewards to $[0, 1]$, but it is an obvious and straight-forward approach. Naturally, the results for the MCTS techniques, which use this reward, will change when a different reward mapping is used, and their results can probably be improved by shaping the reward. In fact, one of the main points of our work is to show that for O-MCTS (as well as for PB-MCTS) no such reward shaping is necessary because these algorithms do not rely on the numerical information. In fact, for them, the mapped linear function with $a \succ b \Leftrightarrow r_{mcts}(a) > r_{mcts}(b)$ is equivalent to the preferences induced by the two-dimensional feedback.

Selected Games. GVGAI provides users with many games. Doing an evaluation on all of them is not feasible. Furthermore, some results would exhibit erratic behavior, since the tested algorithms (except for YOLOBOT) are not suitable for solving some of the games. For example, true rewards often are very sparse, and the agent has to be guided in some way to reliably solve the game.

For this reason, we manually played all the games and selected a variety of interesting, and not too complex games with different characteristics, which we believed to be solvable for the tested algorithms:

- *Zelda*: The agent can hunt monsters and slay them with its sword. It wins by finding the key and taking the door.
- *Chase*: The agent has to catch all animals which flee from the agent. Once an animal finds a caught one, it gets angry and chases the agent. If the agent get caught this way, the game is lost.
- *Whackamole*: The agent can collect mushrooms which spawn randomly. A cat helps it in doing so. The game is won after 2000 time steps or lost if agent and cat collide.
- *Boulderchase*: The agent can dig through sand to a door that opens after it has collected ten diamonds. Monsters chase it through the sand turning sand into diamonds. It may be very hard for a MCTS agent to solve this game.

- *Surround*: The agent can win the game at any time by taking a specific action, or collect points by moving while leaving a snake-like trail. A moving enemy also leaves a trail. The agent shall not collide with trails.
- *Jaws*: The agent controls a submarine, which is hunted by a shark. It can shoot fish giving points and leaving an item behind. Once 20 items are collected, a collision with the shark gives a large number of points, otherwise it loses the game. Colliding with fish always loses the game. The fish spawn randomly on 6 specific positions.
- *Aliens*: The agent can only move from left to right and shoot upwards. Aliens come flying from top to bottom throwing rocks on the agent. For increasing the score, the agent can shoot the aliens or shoot disappearing blocks.

The number of iterations that can be performed by the algorithms depends on the computational budget of calls to the forward model. We tested the algorithms with 250, 500, 1000 and 10000 forward model uses (later called *budget*). Thus, in total, we experimented with 28 problem settings (7 domains × 4 different budgets).

Tuning Algorithms and Experiments. All MCTS algorithms have two parameters in common, the *exploration trade-off* C and *rollout length RL*. For RL we tested 4 different values: $5, 10, 25$ and 50, and for C we tested 9 values from 0 to 2 in steps of size 0.25. In total, these are 36 configurations per algorithm. To reduce variance, we have repeated each experiment 40 times. Overall, 4 algorithms with 36 configurations were run 40 times on 28 problems, resulting in 161280 games played for tuning.

Additionally, we compare the algorithms to YOLOBOT, a highly competitive GVGAI agent that won several challenges [9,13]. YOLOBOT is able to solve games none of the other five algorithms can solve. Note that YOLOBOT is designed and tuned to act within a 20ms time limit. Scaling and even increasing budget might lead to worse and unexpected behavior. Still it is added for sake of comparison and interpretability of strength. For YOLOBOT each of the 28 problems is played 40 times, which leads to 1120 additional games or 162400 games in total.[2]

We are mainly interested on how well the different algorithms perform on the problems, given optimal tuning per problem. To give an answer, we show the performance of the algorithms per problem in percentage of wins and obtained average score. We do a Friedmann test on average ranks of those data with a posthoc Wilcoxon signed rank test to test for significance [6]. Additionally, we show and discuss the performance of all parameter configurations.

4.2 Experiment Results

Table 1 shows the best win rate and the corresponding average score of each algorithm, averaged over 40 runs for each of the 36 different parameter settings. In each row, the best values for the win rate and the average score are shown in

---

[2] Results are available at https://github.com/Muffty/OMCTS_Appendix.

Table 1. The results of algorithms tuned per row.

Game	Time	O-MCTS	MCTS	Yolo-bot	PB-MCTS	MixMax
Jaws	10^4	**100%**	**100%**	27.5%	80%	67.5%
		1083.8	832.7	274.7	895.7	866.8
	10^3	92.5%	**95%**	35%	52.5%	65%
		1028.2	958.9	391.0	788.5	736.4
	500	85%	**90%**	65%	50%	52.5%
		923.4	**1023.1**	705.7	577.6	629.0
	250	**85%**	**85%**	32.5%	37.5%	37.5%
		1000.9	997.6	359.6	548.8	469.0
Surround	10^4	**100%**	**100%**	**100%**	**100%**	**100%**
		81.5	71.0	81.2	64.3	57.6
	10^3	**100%**	**100%**	**100%**	**100%**	**100%**
		83.0	80.8	77.3	40.8	25.0
	500	**100%**	**100%**	**100%**	**100%**	**100%**
		84.6	61.8	83.3	26.3	17.3
	250	**100%**	**100%**	**100%**	**100%**	**100%**
		83.4	64.7	76.1	14.3	10.3
Aliens	10^4	**100%**	**100%**	**100%**	**100%**	**100%**
		82.4	81.6	81.5	81.8	77.0
	10^3	**100%**	**100%**	**100%**	**100%**	**100%**
		79.7	78.4	**82.2**	76.9	76.4
	500	**100%**	**100%**	**100%**	**100%**	**100%**
		78.0	77.3	**81.1**	77.2	76.0
	250	**100%**	**100%**	**100%**	**100%**	**100%**
		77.7	77.1	**79.3**	75.8	74.8
Chase	10^4	**87.5%**	80%	50%	67.5%	37.5%
		6.2	6.0	4.8	5.2	3.9
	10^3	60%	50%	**70%**	30%	17.5%
		4.8	4.8	**5.1**	3.7	2.6
	500	55%	45%	**90%**	27.5%	12.5%
		4.9	4.5	**5.5**	2.9	2.1
	250	40%	32.5%	**90%**	17.5%	7.5%
		4.2	4.1	**5.6**	2.5	2.6
Boulderchase	10^4	62.5%	75%	45%	**82.5%**	30%
		23.7	22.1	18.8	**27.3**	20.1
	10^3	50%	32.5%	**52.5%**	40%	22.5%
		22.8	18.6	21.8	18.1	16.2
	500	**47.5%**	30%	35%	32.5%	15%
		24.7	20.2	18.3	19.4	14.4
	250	40%	40%	**60%**	17.5%	15%
		20.9	20.1	**21.7**	14.7	15.3
Whackamole	10^4	**100%**	**100%**	75%	97.5%	75%
		72.5	44.4	37.0	60.1	48.5
	10^3	**100%**	**100%**	55%	77.5%	65%
		64.0	41.8	33.9	43.9	39.0
	500	**100%**	**100%**	57.5%	70%	52.5%
		59.5	50.0	29.0	38.1	35.4
	250	97.5%	**100%**	50%	65%	52.5%
		54.8	45.9	28.5	35.1	26.6
Zelda	10^4	**97.5%**	87.5%	95%	90%	70%
		8.3	7.4	3.8	**9.6**	8.1
	10^3	80%	85%	**87.5%**	57.5%	42.5%
		8.8	7.5	5.3	8.6	**8.8**
	500	62.5%	75%	**77.5%**	50%	35%
		8.6	8.2	4.6	**8.8**	7.8
	250	55%	55%	**70%**	45%	30%
		8.4	7.8	4.4	8.0	7.2
∅ Rank		1.6	2.5	2.6	3.5	4.7

bold, and a ranking of the algorithms is computed. The resulting average ranks are shown in the last line. We use a Friedmann test and a posthoc Wilcoxon signed rank test as an indication for significant differences in performance. The results of the latter (with a significance level of 99%) are shown in Fig. 3.

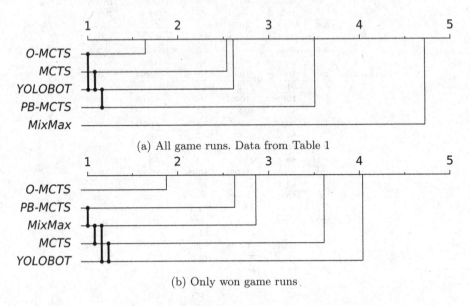

(a) All game runs. Data from Table 1

(b) Only won game runs

Fig. 3. All game runs. Data from Table 1

We can see that O-MCTS performed best with an average rank of 1.6 and a significantly better performance than MCTS and PB-MCTS. Table 1 allows us to take a closer look on the domains. For games that are easy to win, such as *Surround*, *Aliens*, and *Whackamole* O-MCTS beats MCTS and PB-MCTS by winning with a higher score. In *Chase*, a deadly but more deterministic game, O-MCTS is able to achieve a higher win rate. In deadly and stochastic games like *Zelda*, *Boulderchase* and *Jaws* O-MCTS performs comparable to the other algorithms without anyone performing significant better than the others.

Figure 3b summarizes the results when only won games are considered. It can be seen, that in this case, PB-MCTS is significantly better than MCTS. This implies that if PB-MCTS manages to win, it does so with a greater score than MCTS, but it wins less often. YOLOBOT falls behind because it is designed to primarily maximize the win rate, not the score.

Inspecting the performance of MIXMAX it can easily be seen that the hereby added bias towards higher scores often results in a death: Looking at only won games (see Fig. 3b) it achieves a higher rank than MCTS, but overall its performance is significantly worse.

In conclusion, we found evidence that O-MCTS's preference for actions that *maximize win rate* works better than MCTS's tendency to *maximize average performance* for the tested domains.

Table 2. Results for different parameters for all algorithms except of YOLOBOT (Rank 15). In each cell, the overall rank over all games and budgets is shown.

	Exploration-Exploitation								
O-MCTS	**0**	**0.25**	**0.5**	**0.75**	**1**	**1.25**	**1.5**	**1.75**	**2**
5	31	17	9	7	4	1	11	3	13
10	39	12	6	5	14	16	2	10	8
25	45	19	24	35	43	27	29	51	20
50	74	59	46	61	49	55	57	69	56
PB-MCTS	**0**	**0.25**	**0.5**	**0.75**	**1**	**1.25**	**1.5**	**1.75**	**2**
5	63	67	71	72	73	62	68	64	65
10	80	83	89	79	66	82	86	78	77
25	97	94	91	100	101	153	104	103	92
50	151	105	154	152	155	108	148	106	145
MixMax	**0**	**0.25**	**0.5**	**0.75**	**1**	**1.25**	**1.5**	**1.75**	**2**
5	146	147	98	99	96	95	107	102	149
10	162	158	161	163	159	150	157	160	156
25	174	167	173	178	165	170	166	164	172
50	181	169	168	171	176	179	177	175	180
MCTS	**0**	**0.25**	**0.5**	**0.75**	**1**	**1.25**	**1.5**	**1.75**	**2**
5	53	18	23	30	41	22	32	28	40
10	60	21	36	25	26	34	38	37	33
25	76	52	42	44	54	48	50	58	47
50	88	70	75	81	85	84	90	87	93

Parameter Optimization. In Table 2 the overall rank over all parameters for all algorithms are shown. It is clearly visible that a low rollout length RL improves performance and is more important to tune correctly than the exploration-exploitation trade-off C. Since YOLOBOT has no parameters, it is not shown. Except for the extreme case of no exploration ($C = 0$), O-MCTS with $RL = 5$ is better than any other MCTS algorithm. The best configuration is O-MCTS with $RL = 5$ and $C = 1.25$.

Video Demonstrations. For each algorithm and game, we have recorded a video where the agent wins[3]. In those videos it can be seen that O-MCTS frequently plays actions that lead to a higher score, whereas MCTS play more safely—often too cautious and averse to risking any potentially deadly effect.

---

[3] The videos are available at https://bit.ly/2ohbYb3.

5   Conclusion

In this paper we proposed O-MCTS, a modification of MCTS that handles the rewards in an ordinal way: Instead of averaging backpropagated values to obtain a value estimation, it estimates the winning probability of an action using the Borda score. By doing so, the magnitude of distances between different reward signals are disregarded, which can be useful in ordinal domains. In our experiments using the GVGAI framework, we compared O-MCTS to MCTS, PB-MCTS, MixMax and Yolobot, a specialized agent for this domain. Overall, O-MCTS achieved higher win rates and reached higher scores than the other algorithms, confirming that this approach can even be useful in domains where numeric reward information is available.

Acknowledgments. This work was supported by the German Research Foundation (DFG project number FU 580/10). We gratefully acknowledge the use of the Lichtenberg high performance computer of the TU Darmstadt for our experiments.

References

1. Xu, L., Honda, J., Sugiyama, M.: Dueling bandits with qualitative feedback. In: Proceedings of the 33rd AAAI Conference on Artificial Intelligence, pp. 5549–5556 (2019)
2. Ailon, N., Karnin, Z., Joachims, T.: Reducing dueling bandits to cardinal bandits. In: International Conference on Machine Learning, pp. 856–864 (2014)
3. Auer, P., Cesa-Bianchi, N., Fischer, P.: Finite-time analysis of the multiarmed bandit problem. Mach. Learn. **47**(2–3), 235–256 (2002)
4. Black, D.: Partial justification of the Borda count. Public Choice **28**(1), 1–15 (1976)
5. Browne, C.B., et al.: A survey of Monte Carlo tree search methods. IEEE Trans. Comput. Intell. AI Games **4**(1), 1–43 (2012). https://doi.org/10.1109/tciaig.2012.2186810
6. Demšar, J.: Statistical comparisons of classifiers over multiple data sets. J. Mach. Learn. Res. **7**(Jan), 1–30 (2006)
7. Jacobsen, E.J., Greve, R., Togelius, J.: Monte mario: platforming with MCTS. In: Proceedings of the 2014 Annual Conference on Genetic and Evolutionary Computation, pp. 293–300. ACM (2014)
8. Jamieson, K.G., Katariya, S., Deshpande, A., Nowak, R.D.: Sparse dueling bandits. In: AISTATS (2015)
9. Joppen, T., Moneke, M.U., Schröder, N., Wirth, C., Fürnkranz, J.: Informed hybrid game tree search for general video game playing. IEEE Trans. Games **10**(1), 78–90 (2018). https://doi.org/10.1109/TCIAIG.2017.2722235
10. Joppen, T., Wirth, C., Fürnkranz, J.: Preference-based Monte Carlo tree search. In: Proceedings of the 41st German Conference on AI (KI-18) (2018)
11. Khalifa, A., Isaksen, A., Togelius, J., Nealen, A.: Modifying MCTS for human-like general video game playing. In: Proceedings of the 25th International Joint Conference on Artificial Intelligence (IJCAI-16), pp. 2514–2520 (2016)
12. Kocsis, L., Szepesvári, C.: Bandit based Monte-Carlo planning. In: Proceedings of the 17th European Conference on Machine Learning (ECML-06), pp. 282–293 (2006)

13. Perez-Liebana, D., Liu, J., Khalifa, A., Gaina, R.D., Togelius, J., Lucas, S.M.: General video game AI: A multi-track framework for evaluating agents, games and content generation algorithms. arXiv preprint arXiv:1802.10363 (2018)
14. Perez-Liebana, D., Samothrakis, S., Togelius, J., Lucas, S.M., Schaul, T.: General video game AI: competition, challenges and opportunities. In: Proceedings of the 30th AAAI Conference on Artificial Intelligence, pp. 4335–4337 (2016)
15. Puterman, M.L.: Markov Decision Processes: Discrete Stochastic Dynamic Programming, 2nd edn. Wiley, Hoboken (2005)
16. Ramamohan, S.Y., Rajkumar, A., Agarwal, S., Agarwal, S.: Dueling bandits: beyond condorcet winners to general tournament solutions. In: Lee, D.D., Sugiyama, M., Luxburg, U.V., Guyon, I., Garnett, R. (eds.) Advances in Neural Information Processing Systems 29, pp. 1253–1261. Curran Associates, Inc. (2016)
17. Silver, D., et al.: Mastering the game of Go without human knowledge. Nature **550**(7676), 354 (2017)
18. Sprinthall, R.C., Fisk, S.T.: Basic Statistical Analysis. Prentice Hall Englewood Cliffs, NJ (1990)
19. Urvoy, T., Clerot, F., Féraud, R., Naamane, S.: Generic exploration and k-armed voting bandits. In: International Conference on Machine Learning, pp. 91–99 (2013)
20. Vargha, A., Delaney, H.D.: A critique and improvement of the "cl" common language effect size statistics of mcgraw and wong. J. Educ. Behav. Stat. **25**(2), 101–132 (2000). http://www.jstor.org/stable/1165329
21. Weng, P.: Markov decision processes with ordinal rewards: reference point-based preferences. In: Proceedings of the 21st International Conference on Automated Planning and Scheduling (ICAPS-11), ICAPS (2011)
22. Yannakakis, G.N., Cowie, R., Busso, C.: The ordinal nature of emotions. In: Proceedings of the 7th International Conference on Affective Computing and Intelligent Interaction (ACII-17) (2017)
23. Zoghi, M., Whiteson, S., Munos, R., Rijke, M.: Relative upper confidence bound for the k-armed dueling bandit problem. In: Proceedings of the 31st International Conference on Machine Learning (ICML-14), pp. 10–18 (2014). http://proceedings.mlr.press/v32/zoghi14.html

Monte Carlo Game Solver

Tristan Cazenave[✉]

LAMSADE, Université Paris-Dauphine, PSL, CNRS, Paris, France
Tristan.Cazenave@dauphine.psl.eu

Abstract. We present a general algorithm to order moves so as to speedup exact game solvers. It uses online learning of playout policies and Monte Carlo Tree Search. The learned policy and the information in the Monte Carlo tree are used to order moves in game solvers. They improve greatly the solving time for multiple games.

1 Introduction

Monte Carlo Tree Search (MCTS) associated to Deep Reinforcement learning has superhuman results in the most difficult perfect information games (Go, Chess and Shogi) [26]. However little has been done to use this kind of algorithms to exactly solve games. We propose to use MCTS associated to reinforcement learning of policies so as to speedup the resolution of various games.

The paper is organized as follows: the second section deals with related work on games. The third section details the move ordering algorithms for various games. The fourth section gives experimental results for these games.

2 Previous Work

In this section we review the different algorithms that have been used to solve games. We then focus on the $\alpha\beta$ solver. As we improve $\alpha\beta$ with MCTS we show the difference to MCTS Solver. We also expose Depth First Proof Number Search as it has solved multiple games. We finish with a description of online policy learning as the resulting policy is used for our move ordering.

2.1 Solving Games

Iterative Deepening Alpha-Beta associated to a heuristic evaluation function and a transposition table is the standard algorithm for playing games such as Chess and Checkers. Iterative Deepening Alpha-Beta has also been used to solve games such as small board Go [30], Renju [29], Amazons endgames [15]. Other researchers have instead used a Depth-first Alpha-Beta without Iterative Deepening and with domain specific algorithms to solve Domineering [28] and Atarigo [3]. The advantage of Iterative Deepening associated to a transposition table for solving games is that it finds the shortest win and that it reuses the information of previous iterations for ordering the moves, thus maximizing the cuts. The

T. Cazenave et al. (Eds.): MCS 2020, CCIS 1379, pp. 56–70, 2021.
https://doi.org/10.1007/978-3-030-89453-5_5

heuristics usually used to order moves in combination with Iterative Deepening are: trying first the transposition table move, then trying killer moves and then sorting the remaining moves according to the History Heuristic [23]. The advantage of not using Iterative Deepening is that the iterations before the last one are not performed, saving memory and time, however if bad choices on move ordering happen, the search can waste time in useless parts of the search tree and can also find move sequences longer than necessary.

There are various competing algorithms for solving games [12]. The most simple are Alpha-Beta and Iterative Deepening Alpha-Beta. Other algorithms memorize the search tree in memory and expand it with a best first strategy: Proof Number Search [2], PN^2 [4], Depth-first Proof Number Search (Df-pn) [17], Monte Carlo Tree Search Solver [8,32] and Product Propagation [20].

Games solved with a best first search algorithm include Go-Moku with Proof Number search and Threat Space Search [1], Checkers using various algorithms [24], Fanorona with PN^2 [22], 6×6 Lines of Action with PN^2 [31], 6×5 Breakthrough with parallel PN^2 [21], and 9×9 Hex with parallel Df-pn [18].

Other games such as Awari were solved using retrograde analysis [19]. Note that retrograde analysis was combined with search to solve Checkers and Fanorona.

2.2 $\alpha\beta$ Solver

Iterative Deepening $\alpha\beta$ has long been the best algorithm for multiple games. Most of the Chess engines still use it even if the current best algorithm is MCTS [26].

Depth first $\alpha\beta$ is more simple but it can be better than Iterative Deepening $\alpha\beta$ for solving games since it does not have to explore a large tree before searching the next depth. More over in the case of games with only two outcomes the results are always either Won or Lost and enable immediate cuts when Iterative Deepening $\alpha\beta$ has to deal with unknown values when it reaches depth zero and the state is not terminal.

One interesting property of $\alpha\beta$ is that selection sort becomes an interesting sorting algorithm. It is often useful to only try the best move or a few good moves before reaching a cut. Therefore it is not necessary to sort all the moves at first. Selecting move by move as in selection sorting can be more effective.

2.3   MCTS Solver

MCTS has already been used as a game solver [32]. The principle is to mark as solved the subtrees that have an exact result. As the method uses playouts it has to go through the tree at each playout and it revisits many times the same states doing costly calculations to choose the move to try according to the bandit. Moreover in order to solve a game a large game tree has to be kept in memory.

The work on MCTS Solver has later been extended to games with multiple outcomes [8].

Algorithm 1. The $\alpha\beta$ algorithm for solving games.

```
 1: Function αβ (s,depth,α,β)
 2:    if isTerminal (s) or depth = 0 then
 3:        return Evaluation (s)
 4:    end if
 5:    if s has an entry t in the transposition table then
 6:        if the result of t is exact then
 7:            return t.res
 8:        end if
 9:        put t.move as the first legal move
10:    end if
11:    for move in legal moves for s do
12:        s₁ = play (s, move)
13:        eval = -αβ(s₁,depth − 1,-β,-α)
14:        if eval > α then
15:            α = eval
16:        end if
17:        if α ≥ β then
18:            update the transposition table
19:            return β
20:        end if
21:    end for
22:    update the transposition table
23:    return α
```

2.4 Depth First Proof Number Search

Proof Number Search is a best first algorithm that keeps the search tree in memory so as to expand the most informative leaf [2]. In order to solve the memory problem of Proof Number Search, the PN^2 algorithm has been used [4]. PN^2 uses a secondary Proof Number Search at each leaf of the main Proof Number Search tree, thus enabling the square of the total number of nodes of the main search tree to be explored. More recent developments of Proof Number Search focus on Depth-First Proof Number search (DFPN) [17]. The principle is to use a transposition table and recursive depth first search to efficiently search the game tree and solve the memory problems of Proof Number Search. DFPN can be parallelized to improve the solving time [13]. It has been improved for the game of Hex using a trained neural network [10]. It can be applied to many problems, including recently Chemical Synthesis Planning [14].

2.5 Online Policy Learning

Playout Policy Adaptation with Move Features (PPAF) has been applied to many games [7].

An important detail of the playout algorithm is the code function. In PPAF the same move can have different codes that depend on the presence of features associated to the move. For example in Breakthrough the code also takes into account whether the arriving square is empty or contains an opponent pawn.

The principle of the learning algorithm is to add 1.0 to the weight of the moves played by the winner of the playout. It also decreases the weights of the moves not played by the winner of the playout by a value proportional to the exponential of the weight. This algorithm is given in Algorithm 2.

Algorithm 2. The PPAF adapt algorithm

1: Function adapt $(winner, board, player, playout, policy)$
2:   $polp \leftarrow policy$
3:   **for** $move$ in $playout$ **do**
4:     **if** $winner = player$ **then**
5:       $polp$ [code$(move)$] $\leftarrow polp$ [code$(move)$] $+ \alpha$
6:       $z \leftarrow 0.0$
7:       **for** m in possible moves on $board$ **do**
8:         $z \leftarrow z + \exp (policy$ [code(m)]$)$
9:       **end for**
10:       **for** m in possible moves on $board$ **do**
11:         $polp$ [code(m)] $\leftarrow polp$ [code(m)] $- \alpha * \frac{exp(policy[code(m)])}{z}$
12:       **end for**
13:     **end if**
14:     play $(board, move)$
15:     $player \leftarrow$ opponent $(player)$
16:   **end for**
17:   $policy \leftarrow polp$

2.6 GRAVE

The principle of the All Moves As First heuristic (AMAF) is to compute statistics where all the moves of the playout are taken into account for the average of the moves. The principle of RAVE is to start from AMAF statistics when there are only a few playouts because the AMAF statistics are then more precise than the UCT statistics and to progressively switch to UCT statistics when more playouts are available. The formula found using a mathematical analysis [11] is:

$$\beta_m \leftarrow \frac{pAMAF_m}{pAMAF_m + p_m + bias \times pAMAF_m \times p_m}$$

$$argmax_m((1.0 - \beta_m) \times mean_m + \beta_m \times AMAF_m)$$

Where $pAMAF_m$ is the number of playouts that contain move m, p_m is the number of playouts that start with move m, $bias$ is a parameter to be tuned, $mean_m$ is the average of the playouts that start with move m and $AMAF_m$ is the average of the playouts that contain move m.

GRAVE is a simple but effective improvement of RAVE. GRAVE takes into account the AMAF statistics of the last state in the tree that has been visited more than a fixed number of times instead of always taking into account the AMAF statistics of the current state. In this way the states with few playouts use more meaningful AMAF statistics. GRAVE has very good results in General Game Playing [5,27].

3   Move Ordering for Different Games

We describe the general tools used for move ordering then their adaptation to different games.

3.1   Outline

In order to collect useful information to order moves we use a combination of the GRAVE algorithm [6] and of the PPAF algorithm. Once the Monte Carlo search with GRAVE and PPAF is finished we use the transposition table of the Monte Carlo search to order the moves, putting first the most simulated ones. When outside the transposition table we use the weights learned by PPAF to order the moves. The algorithm used to score the moves so as to order them is given in Algorithm 3.

Algorithm 3. The Monte Carlo Move Ordering function

1: Function orderMC ($board$, $code$)
2:     $score \leftarrow policy[code]$
3:     **if** $board$ has an entry t in the MCTS TT **then**
4:         **if** $t.nbPlayouts > 100$ **then**
5:             **for** $move$ in legal moves for $board$ **do**
6:                 **if** $t.nbPlayouts[move] > 0$ **then**
7:                     **if** $code(move) = code$ **then**
8:                         $score \leftarrow t.nbPlayouts[move]$
9:                     **end if**
10:                 **end if**
11:             **end for**
12:         **end if**
13:     **end if**
14:     **return** $1000000000 - 1000 \times score$

3.2   Atarigo

Atarigo is a simplification of the game of Go. The first player to capture has won. It is a game often used to teach Go to beginners. Still it is an interesting games and tactics can be hard to master.

  The algorithm for move ordering is given in Algorithm 4. It always puts first a capture move since it wins the game. If no such move exist it always plays a move that saves a one liberty string since it is a forced move to avoid losing. Then it favors moves on liberties of opponent strings that have few liberties provided the move has itself sufficient liberties. If none of these are available it returns the evaluation by the Monte Carlo ordering function.

The code associated to a move is calculated using the colors of the four intersections next to the move.

Algorithm 4. The function to order moves at Atarigo

```
1:  Function order (board, move)
2:     minOrder ← 361
3:     for i in adjacents to move do
4:        if i is an opponent stone then
5:           n ← number of liberties of i
6:           if n = 1 then
7:              return 0
8:           end if
9:           nb ← n − 4 × nbEmptyAdjacent(move)
10:          if nb < minOrder then
11:             minOrder ← nb
12:          end if
13:       end if
14:    end for
15:    if move escapes an atari then
16:       return 1
17:    end if
18:    if minOrder = 361 then
19:       if MonteCarloMoveOrdering then
20:          return orderMC(board, code(move))
21:       end if
22:       return 20 − nbEmptyAdjacent(move)
23:    end if
24:    return minOrder
```

3.3 Nogo

Nogo is the misere version of Atarigo [9]. It was introduced at the 2011 Combinatorial Game Theory Workshop in Banff. The first player to capture has lost. It is usually played on small boards. In Banff there was a tournament for programs and Bob Hearn won the tournament using Monte-Carlo Tree Search.

We did not find simple heuristics to order moves at Nogo. So the standard algorithm uses no heuristic and the MC algorithms sort moves according to Algorithm 3.

3.4 Go

Go was solved for rectangular boards up to size 7×4 by the MIGOS program [30]. The algorithm used was an iterative deepening $\alpha\beta$ with transposition table. We use no heuristic to sort moves at Go and completely rely on Algorithm 3 to order moves.

3.5   Breakthrough and Knightthrough

Breakthrough is an abstract strategy board game invented by Dan Troyka in 2000. It won the 2001 8×8 Game Design Competition and it is played on Little Golem. The game starts with two rows of pawns on each side of the board. Pawns can capture diagonally and go forward either vertically or diagonally. The first player to reach the opposite row has won. Breakthrough has been solved up to size 6×5 using Job Level Proof Number Search [21]. The code for a move at Breakthrough contains the starting square, the arrival square and whether it is empty or contains an enemy pawn. The ordering gives priority to winning moves, then to moves to prevent a loss, then Monte Carlo Move Ordering.

Misere Breakthrough is the misere version of Breakthrough, the games is lost if a pawn reaches the opposite side. It is also a difficult game and its is more difficult for MCTS algorithms [7]. The code for a move is the same as for Breakthrough and the ordering is Monte Carlo Move Ordering.

Knightthrough emerged as a game invented for the General Game Playing competitions. Pawns are replaced with knights. Misere Knightthrough is the misere version of the game where the goal is to lose. Codes for moves and move ordering are similar to Breakthrough.

3.6   Domineering

Domineering is played on a chess board and two players alternate putting domi-noes on the board. The first player puts the dominoes vertically, the second player puts them horizontally. The first player who cannot play loses. In Misere Domineering the first player who cannot play wins. We use no heuristic to sort moves at Domineering and completely rely on Algorithm 3 to order moves.

Algorithm 5. The function to order moves at Knightthrough

```
 1: Function order (board, move)
 2:    if move is a winning move then
 3:       return 0
 4:    end if
 5:    if move captures an opponent piece then
 6:       if capture in the first 3 lines then
 7:          return 1
 8:       end if
 9:    end if
10:    if destination in the last 3 lines then
11:       if support(destination) > attack(destination) then
12:          return 2
13:       end if
14:    end if
15:    if MonteCarloMoveOrdering then
16:       return orderMC(board, code(move))
17:    end if
18:    return 100
```
---

4 Experimental Results

The iterative deepening $\alpha\beta$ with a transposition table (ID $\alpha\beta$ TT) is called with a null window since it saves much time compared to calling it with a full window. Other algorithms are called with the full window since they only deal with terminal states values and that the games we solve are either Won or Lost.

A transposition table containing 1 048 575 entries is used for all games. An entry in the transposition table is always replaced by a new one.

An algorithm name finishing with MC denotes the use of Monte Carlo Move Ordering. The times given for MC algorithms include the time for the initial MCTS that learns a policy. The original Proof Number Search algorithm is not included in the experiments since it fails due to being short of memory for complex games. The PN^2 algorithm solves this memory problem and is included in the experiments. The algorithms that do not use MC still do some move ordering but without Monte Carlo. For example in Algorithm 4 for Atarigo the MonteCarloMoveOrdering boolean is set to False but the function to order moves is still used.

Table 1 gives the results for Atarigo. For Atarigo 5×5 $\alpha\beta$ TT MC is the best algorithm and is much better than $\alpha\beta$ TT. For Atarigo 6×5 the best algorithm is again $\alpha\beta$ TT MC which is much better than all other algorithms.

Table 2 gives the results for Nogo. Nogo 7×3 is solved in 49.72 s by $\alpha\beta$ TT MC with 100 000 playouts. This is 88 times faster than $\alpha\beta$ TT the best algorithm not using MC.

Table 1. Different algorithms for solving Atarigo.

Size	5×5	
Result	Won	
	Move count	Time
PN^2	14 784 088 742	37 901.56 s.
ID $\alpha\beta$ TT	> 35 540 000 000	> 86 400.00 s.
$\alpha\beta$ TT	> 37 660 000 000	> 86 400.00 s.
ID $\alpha\beta$ TT MC	62 800 334	126.84 s.
$\alpha\beta$ TT MC	**3 956 049**	**12.79 s.**

Size	6×5	
Result	Won	
	Move count	Time
PN^2	> 33 150 000 000	> 86 400.00 s.
ID $\alpha\beta$ TT	> 37 190 000 000	> 86 400.00 s.
$\alpha\beta$ TT	> 7 090 000 000	> 44 505.91 s.
ID $\alpha\beta$ TT MC	12 713 931 627	27 298.35 s.
$\alpha\beta$ TT MC	**329 780 434**	**787.17 s.**

Table 2. Different algorithms for solving Nogo.

Size	7×3	
Result	Won	
	Move count	Time
PN^2	> 80 390 000 000	> 86 400.00 s.
ID $\alpha\beta$ TT	10 921 978 839	12 261.64 s.
$\alpha\beta$ TT	3 742 927 598	4 412.21 s.
ID $\alpha\beta$ TT MC	1 927 635 856	2 648.91 s.
$\alpha\beta$ TT MC	**35 178 886**	**49.72 s.**

Size	5×4	
Result	Won	
	Move count	Time
PN^2	> 101 140 000 000	> 86 400.00 s.
ID $\alpha\beta$ TT	1 394 182 870	1 573.72 s.
$\alpha\beta$ TT	1 446 922 704	1 675.64 s.
ID $\alpha\beta$ TT MC	73 387 083	134.26 s.
$\alpha\beta$ TT MC	**33 850 535**	**74.77 s.**

Nogo 5×4 is solved best by $\alpha\beta$ TT MC with 1 000 000 playouts before the $\alpha\beta$ search. It is 21 times faster than ID $\alpha\beta$ TT the best algorithm not using MC.

$\alpha\beta$ TT MC with 10 000 000 playouts solves Nogo 5×5 in 61 430.88 s and 46 092 056 485 moves. Nogo 5×5 was first solved in 2013 [25]. The solution we found is given in Fig. 1.

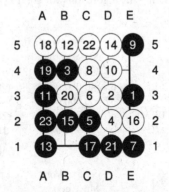

Fig. 1. Solution of Nogo 5×5.

As it is the first time results about solving Nogo are given we recapitulate in Table 3 the winner for various sizes. A one means a first player win and a two a second player win.

Table 3. Winner for Nogo boards of various sizes

	1	2	3	4	5	6	7	8	9	10
1	2	1	1	2	1	1	1	1	1	1
2	1	1	2	2	1	1	1	1	2	2
3	1	2	1	2	1	1	1	1		
4	2	2	2	2	1	1				
5	1	1	1	1	1					
6	1	1	1	1						
7	1	1	1							
8	1	1	1							
9	1	2								
10	1	2								

Table 4. Different algorithms for solving Go.

Size	3×3	
Result	Won	
	Move count	Time
PN^2	246 394	3.72 s.
ID $\alpha\beta$ TT	840 707	11.34 s.
$\alpha\beta$ TT	420 265	11.50 s.
ID $\alpha\beta$ TT MC	375 414	5.62 s.
$\alpha\beta$ TT MC	**6 104**	**0.16 s.**

Size	4×3	
Result	Won	
	Move count	Time
PN^2	43 202 038	619.98 s.
ID $\alpha\beta$ TT	39 590 950	515.71 s.
$\alpha\beta$ TT	107 815 563	1 977.86 s.
ID $\alpha\beta$ TT MC	22 382 730	348.08 s.
$\alpha\beta$ TT MC	**4 296 893**	**96.63 s.**

Table 4 gives the results for Go. Playouts and depth first $\alpha\beta$ can last a very long time in Go since stones are captured and if random play occurs the goban can become almost empty again a number of times before the superko rules forbids states. In order to avoid very long and useless games an artificial limit on the number of moves allowed in a game was set to twice the size of the board. This is not entirely satisfactory since one can imagine weird cases where the limit is not enough. The problem does not appear in the other games we have solved since they converge to a terminal state before a fixed number of moves. The trick we use to address the problem is to send back an evaluation of zero if the search reaches the limit. When searching for a win with a null window this is equivalent to a loss and when searching for a loss it is equivalent to a win. Therefore if the search finds a win it does not rely on the problematic states. The 3×3 board was solved with a komi of 8.5, the 4×3 board was solved with a komi of 3.5.

Table 5. Different algorithms for solving Breakthrough.

Size	5×5	
Result	Lost	
	Move count	Time
PN^2	> 38 780 000 000	> 86 400.00 s.
ID $\alpha\beta$ TT	13 083 392 799	33 590.59 s.
$\alpha\beta$ TT	19 163 127 770	43 406.79 s.
ID $\alpha\beta$ TT MC	3 866 853 361	11 319.39 s.
$\alpha\beta$ TT MC	**3 499 173 137**	**9 243.66 s.**

Table 5 gives the results for Breakthrough. Using MC improves much the solving time. $\alpha\beta$ with MC uses seven times less nodes than the previous algorithm that solved Breakthrough 5×5 without patterns (i.e. parallel PN^2 with 64 clients [21]). Using endgame patterns divides by seven the number of required nodes for parallel PN^2.

Table 6. Different algorithms for solving Misere Breakthrough.

Size	4×5	
Result	Lost	
	Move count	Time
PN^2	> 42 630 000 000	> 86 400 s.
ID $\alpha\beta$ TT	> 43 350 000 000	> 86 400 s.
$\alpha\beta$ TT	> 42 910 000 000	> 86 400 s.
ID $\alpha\beta$ TT MC	1 540 153 635	3 661.50 s.
$\alpha\beta$ TT MC	**447 879 697**	**1 055.32 s.**

Table 6 gives the results for Misere Breakthrough. $\alpha\beta$ TT MC is the best algorithm and is much better than all non MC algorithms.

Table 7. Different algorithms for solving Knightthrough.

Size	6×6	
Result	Won	
	Move count	Time
PN^2	> 33 110 000 000	> 86 400 s.
ID $\alpha\beta$ TT	1 153 730 169	4 894.69 s.
$\alpha\beta$ TT	2 284 038 427	6 541.08 s.
ID $\alpha\beta$ TT MC	**17 747 503**	**102.60 s.**
$\alpha\beta$ TT MC	528 783 129	1 699.01 s.

Size	7×6	
Result	Won	
	Move count	Time
PN^2	> 30 090 000 000	> 86 400 s.
ID $\alpha\beta$ TT	> 17 500 000 000	> 86 400 s.
$\alpha\beta$ TT	> 29 980 000 000	> 86 400 s.
ID $\alpha\beta$ TT MC	**2 540 383 012**	**13 716.36 s.**
$\alpha\beta$ TT MC	6 650 804 159	23 958.04 s.

The results for Knightthrough are in Table 7. ID $\alpha\beta$ TT MC is the best algorithm and far better than algorithms not using MC. This is the first time Knightthrough 7×6 is solved.

Table 8. Different algorithms for solving Misere Knightthrough.

Size	5×5	
Result	Lost	
	Move count	Time
PN^2	> 45 290 000 000	> 86 400 s.
ID $\alpha\beta$ TT	> 52 640 000 000	> 86 400 s.
$\alpha\beta$ TT	> 56 230 000 000	> 86 400 s.
ID $\alpha\beta$ TT MC	> 41 840 000 000	> 86 400 s.
$\alpha\beta$ TT MC	**20 375 687 163**	**42 425.41 s.**

Table 8 gives the results for Misere Knightthrough. Misere Knightthrough 5×5 is solved in 20 375 687 163 moves and 42 425.41 s by $\alpha\beta$ TT MC. This is the first time Misere Knightthrough 5×5 is solved. Misere Knightthrough 5×5 is much more difficult to solve than Knightthrough 5×5 which is solved in seconds by ID $\alpha\beta$ TT MC. This is due to Misere Knightthrough being a waiting game with longer games than Knightthrough.

Table 9. Different algorithms for solving Domineering.

Size	7×7	
Result	Won	
	Move count	Time
PN^2	> 41 270 000 000	> 86 400 s.
ID $\alpha\beta$ TT	18 958 604 687	35 196.62 s.
$\alpha\beta$ TT	197 471 137	376.23 s.
ID $\alpha\beta$ TT MC	2 342 641 133	5 282.06 s.
$\alpha\beta$ TT MC	**29 803 373**	**123.76 s.**

Table 9 gives the results for Domineering. The best algorithm is $\alpha\beta$ TT MC which is 3 times faster than $\alpha\beta$ TT without MC.

Table 10 gives the results for Misere Domineering. The best algorithm is $\alpha\beta$ TT MC which is much better than all non MC algorithms.

Table 10. Different algorithms for solving Misere Domineering.

Size	7×7	
Result	Won	
	Move count	Time
PN^2	> 44 560 000 000	> 86 400 s.
ID $\alpha\beta$ TT	> 49 290 000 000	> 86 400 s.
$\alpha\beta$ TT	> 49 580 000 000	> 86 400 s.
ID $\alpha\beta$ TT MC	7 013 298 932	14 936.03 s.
$\alpha\beta$ TT MC	**72 728 678**	**212.25 s.**

5   Conclusion

For the games we solved, Misere Games are more difficult to solve than normal games. In Misere Games the player waits and tries to force the opponent to play a losing move. This makes the game longer and reduces the number of winning sequences and winning moves.

Monte Carlo Move Ordering improves much the speed of $\alpha\beta$ with transposition table compare to depth first $\alpha\beta$ and Iterative Deepening $\alpha\beta$ with transposition table but without Monte Carlo Move Ordering. The experimental results show significant improvements for nine different games.

In future work we plan to parallelize the algorithms and apply them to other problems. It would also be interesting to test if improved move ordering due to Monte Carlo Move Ordering would improve other popular solving algorithms such as DFPN. The ultimate goal with this kind of algorithms could be to solve exactly the game of Chess which is possible provided we have a very strong move ordering algorithm [16].

Acknowledgment. This work was supported in part by the French government under management of Agence Nationale de la Recherche as part of the "Investissements d'avenir" program, reference ANR19-P3IA-0001 (PRAIRIE 3IA Institute).

References

1. Allis, L.V., van den Herik, H.J., Huntjens, M.P.H.: Go-Moku solved by new search techniques. Comput. Intell. **12**, 7–23 (1996)
2. Allis, L.V., van der Meulen, M., van den Herik, H.J.: Proof-number search. Artif. Intell. **66**(1), 91–124 (1994)
3. Boissac, F., Cazenave, T.: De nouvelles heuristiques de recherche appliquées à la résolution d'Atarigo. In: Intelligence artificielle et jeux, pp. 127–141. Hermes Science (2006)
4. Breuker, D.M.: Memory versus search in games. Ph.D. thesis, Universiteit Maastricht (1998)
5. Browne, C., Stephenson, M., Piette, É., Soemers, D.J.: A practical introduction to the Ludii general game system. In: Cazenave, T., van den Herik, J., Saffidine, A., Wu, I.C. (eds.) ACG 2019. LNCS, vol. 12516, pp. 167–179. Springer, Cham (2020). https://doi.org/10.1007/978-3-030-65883-0_14
6. Cazenave, T.: Generalized rapid action value estimation. In: IJCAI 2015, pp. 754–760 (2015)
7. Cazenave, T.: Playout policy adaptation with move features. Theor. Comput. Sci. **644**, 43–52 (2016)
8. Cazenave, T., Saffidine, A.: Score bounded Monte-Carlo tree search. In: van den Herik, H.J., Iida, H., Plaat, A. (eds.) CG 2010. LNCS, vol. 6515, pp. 93–104. Springer, Heidelberg (2011). https://doi.org/10.1007/978-3-642-17928-0_9
9. Chou, C.-W., Teytaud, O., Yen, S.-J.: Revisiting Monte-Carlo tree search on a normal form game: NoGo. In: Di Chio, C., et al. (eds.) EvoApplications 2011. LNCS, vol. 6624, pp. 73–82. Springer, Heidelberg (2011). https://doi.org/10.1007/978-3-642-20525-5_8
10. Gao, C., Müller, M., Hayward, R.: Focused depth-first proof number search using convolutional neural networks for the game of hex. In: IJCAI 2017, pp. 3668–3674 (2017)
11. Gelly, S., Silver, D.: Monte-Carlo tree search and rapid action value estimation in computer go. Artif. Intell. **175**(11), 1856–1875 (2011)
12. van den Herik, H.J., Uiterwijk, J.W.H.M., van Rijswijck, J.: Games solved: now and in the future. Artif. Intell. **134**(1–2), 277–311 (2002)
13. Hoki, K., Kaneko, T., Kishimoto, A., Ito, T.: Parallel dovetailing and its application to depth-first proof-number search. ICGA J. **36**(1), 22–36 (2013)
14. Kishimoto, A., Buesser, B., Chen, B., Botea, A.: Depth-first proof-number search with heuristic edge cost and application to chemical synthesis planning. In: Advances in Neural Information Processing Systems, pp. 7224–7234 (2019)
15. Kloetzer, J., Iida, H., Bouzy, B.: A comparative study of solvers in amazons endgames. In: IEEE Symposium On Computational Intelligence and Games, CIG 2008, pp. 378–384. IEEE (2008)
16. Lemoine, J., Viennot, S.: Il n'est pas impossible de résoudre le jeu d'échecs. 1024 - Bulletin de la société informatique de France 6, Juillet 2015
17. Nagai, A.: DF-PN algorithm for searching AND/OR trees and its applications. Ph.D. thesis, The University of Tokyo (2002)

18. Pawlewicz, J., Hayward, R.B.: Scalable parallel DFPN search. In: van den Herik, H.J., Iida, H., Plaat, A. (eds.) CG 2013. LNCS, vol. 8427, pp. 138–150. Springer, Cham (2014). https://doi.org/10.1007/978-3-319-09165-5_12

19. Romein, J.W., Bal, H.E.: Solving awari with parallel retrograde analysis. IEEE Comput. **36**(10), 26–33 (2003)

20. Saffidine, A., Cazenave, T.: Developments on product propagation. In: van den Herik, H.J., Iida, H., Plaat, A. (eds.) CG 2013. LNCS, vol. 8427, pp. 100–109. Springer, Cham (2014). https://doi.org/10.1007/978-3-319-09165-5_9

21. Saffidine, A., Jouandeau, N., Cazenave, T.: Solving BREAKTHROUGH with race patterns and job-level proof number search. In: van den Herik, H.J., Plaat, A. (eds.) ACG 2011. LNCS, vol. 7168, pp. 196–207. Springer, Heidelberg (2012). https://doi.org/10.1007/978-3-642-31866-5_17

22. Schadd, M.P.D., Winands, M.H.M., Uiterwijk, J.W.H.M., van den Herik, H.J., Bergsma, M.H.J.: Best play in Fanorona leads to draw. New Math. Nat. Comput. **4**(03), 369–387 (2008)

23. Schaeffer, J.: The history heuristic and alpha-beta search enhancements in practice. IEEE Trans. Pattern Anal. Mach. Intell. **11**(11), 1203–1212 (1989)

24. Schaeffer, J., et al.: Checkers is solved. Science **317**(5844), 1518–1522 (2007)

25. She, P.: The design and study of NoGo program. Master's thesis, National Chiao Tung University, Taiwan (2013)

26. Silver, D., et al.: A general reinforcement learning algorithm that masters chess, shogi, and go through self-play. Science **362**(6419), 1140–1144 (2018)

27. Sironi, C.F.: Monte-Carlo tree search for artificial general intelligence in games. Ph.D. thesis, Maastricht University (2019)

28. Uiterwijk, J.W.H.M.: 11 × 11 domineering is solved: the first player wins. In: Plaat, A., Kosters, W., van den Herik, J. (eds.) CG 2016. LNCS, vol. 10068, pp. 129–136. Springer, Cham (2016). https://doi.org/10.1007/978-3-319-50935-8_12

29. Wágner, J., Virág, I.: Solving renju. ICGA J. **24**(1), 30–35 (2001)

30. van der Werf, E.C., Winands, M.H.: Solving go for rectangular boards. ICGA J. **32**(2), 77–88 (2009)

31. Winands, M.H.: 6 × 6 LOA is solved. ICGA J. **31**(4), 234–238 (2008)

32. Winands, M.H.M., Björnsson, Y., Saito, J.-T.: Monte-Carlo tree search solver. In: van den Herik, H.J., Xu, X., Ma, Z., Winands, M.H.M. (eds.) CG 2008. LNCS, vol. 5131, pp. 25–36. Springer, Heidelberg (2008). https://doi.org/10.1007/978-3-540-87608-3_3

Generalized Nested Rollout Policy Adaptation

Tristan Cazenave[✉]

LAMSADE, Université Paris-Dauphine, PSL, CNRS, Paris, France
`Tristan.Cazenave@dauphine.psl.eu`

Abstract. Nested Rollout Policy Adaptation (NRPA) is a Monte Carlo search algorithm for single player games. In this paper we propose to generalize NRPA with a temperature and a bias and to analyze theoretically the algorithms. The generalized algorithm is named GNRPA. Experiments show it improves on NRPA for different application domains: SameGame and the Traveling Salesman Problem with Time Windows.

1 Introduction

Monte Carlo Tree Search (MCTS) has been successfully applied to many games and problems [4].

Nested Monte Carlo Search (NMCS) [5] is an algorithm that works well for puzzles and optimization problems. It biases its playouts using lower level playouts. At level zero NMCS adopts a uniform random playout policy. Online learning of playout strategies combined with NMCS has given good results on optimization problems [22]. Other applications of NMCS include Single Player General Game Playing [16], Cooperative Pathfinding [2], Software testing [20], heuristic Model-Checking [21], the Pancake problem [3], Games [8] and the RNA inverse folding problem [18].

Online learning of a playout policy in the context of nested searches has been further developed for puzzles and optimization with Nested Rollout Policy Adaptation (NRPA) [23]. NRPA has found new world records in Morpion Solitaire and crosswords puzzles. NRPA has been applied to multiple problems: the Traveling Salesman with Time Windows (TSPTW) problem [9,11], 3D Packing with Object Orientation [13], the physical traveling salesman problem [14], the Multiple Sequence Alignment problem [15] or Logistics [12]. The principle of NRPA is to adapt the playout policy so as to learn the best sequence of moves found so far at each level.

The use of Gibbs sampling in Monte Carlo Tree Search dates back to the general game player Cadia Player and its MAST playout policy [1].

We now give the outline of the paper. The second section describes NRPA. The third section gives a theoretical analysis of NRPA. The fourth section describes the generalization of NRPA. The fifth section details optimizations of GNRPA. The sixth section gives experimental results for SameGame and TSPTW.

© Springer Nature Switzerland AG 2021
T. Cazenave et al. (Eds.): MCS 2020, CCIS 1379, pp. 71–83, 2021.
https://doi.org/10.1007/978-3-030-89453-5_6

2   NRPA

NRPA learns a rollout policy by adapting weights on each action. Vanilla NRPA starts with all weights set to zero. During the playout phase, action is sampled with a probability proportional to the exponential of the associated weight. The playout algorithm is given in Algorithm 1. The algorithm starts with initializing the sequence of moves that it will play (line 2). Then it performs a loop until it reaches a terminal states (lines 3–6). At each step of the playout it calculates the sum of all the exponentials of the weights of the possible moves (lines 7–10) and chooses a move proportional to its probability given by the softmax function (line 11). Then it plays the chosen move and adds it to the sequence of moves (lines 12–13). Each move is associated to a code which is usually independent of the state.

Then, the policy is adapted on the best current sequence found, by increasing the weight of the best actions and decreasing the weights of all the moves proportionally to their probabilities of being played. The Adapt algorithm is given in Algorithm 2. For all the states of the sequence passed as a parameter it adds α to the weight of the move of the sequence (lines 3–5). Then it reduces all the moves proportionally to $\alpha \times$ the probability of playing the move so as to keep the sum of logits unchanged (lines 6–12).

In NRPA, each nested level takes as input a policy, and returns a sequence. At each step, the algorithm makes a recursive call to the lower level and gets a sequence as a result. It adapts the policy to the best sequence of the level at each step. At level zero it makes a playout.

The NRPA algorithm is given in Algorithm 3. At level zero it simply performs a playout (lines 2–3). At greater levels it performs N iterations and for each iteration it calls itself recursively to get a score and a sequence (lines 4–7). If it finds a new best sequence for the level it keeps it as the best sequence (lines 8–11). Then it adapts the policy using the best sequence found so far at the current level (line 12).

NRPA balances exploitation by adapting the probabilities of playing moves toward the best sequence of the level, and exploration by using Gibbs sampling

---

Algorithm 1. The playout algorithm

1: playout (*state*, *policy*)
2:     *sequence* ← []
3:     **while** true **do**
4:         **if** *state* is terminal **then**
5:             **return** (score (*state*), *sequence*)
6:         **end if**
7:         $z \leftarrow 0.0$
8:         **for** $m \in$ possible moves for *state* **do**
9:             $z \leftarrow z + \exp (policy \, [\text{code}(m)])$
10:        **end for**
11:        choose a *move* with probability $\frac{exp(policy[code(move)])}{z}$
12:        *state* ← play (*state*, *move*)
13:        *sequence* ← *sequence* + *move*
14:    **end while**

Algorithm 2. The Adapt algorithm

1: Adapt (*policy, sequence*)
2:    *polp ← policy*
3:    *state ← root*
4:    **for** *move ∈ sequence* **do**
5:        *polp* [code(*move*)] *← polp* [code(*move*)] + α
6:        $z ← 0.0$
7:        **for** $m ∈$ possible moves for *state* **do**
8:            $z ← z +$ exp (*policy* [code(m)])
9:        **end for**
10:       **for** $m ∈$ possible moves for *state* **do**
11:           *polp* [code(m)] *← polp* [code(m)] - $\alpha * \frac{exp(policy[code(m)])}{z}$
12:       **end for**
13:       *state ←* play (*state, move*)
14:   **end for**
15:   *policy ← polp*

Algorithm 3. The NRPA algorithm.

1: NRPA (*level, policy*)
2:    **if** level == 0 **then**
3:        **return** playout (root, *policy*)
4:    **else**
5:        *bestScore ← −∞*
6:        **for** N iterations **do**
7:            (result,new) *←* NRPA(*level* − 1, *policy*)
8:            **if** result ≥ bestScore **then**
9:                bestScore *←* result
10:               seq *←* new
11:           **end if**
12:           policy *←* Adapt (policy, seq)
13:       **end for**
14:       **return** (bestScore, seq)
15:   **end if**

at the lowest level. It is a general algorithm that has proven to work well for many optimization problems.

3   Theoretical Analysis of NRPA

In NRPA each move is associated to a weight. The goal of the algorithm is to learn these weights so as to produce a playout policy that generates good sequences of moves. At each level of the algorithm the best sequence found so far is memorized. Let $s_1, ..., s_m$ be the sequence of states of the best sequence. Let n_i be the number of possible moves in a state s_i. Let $m_{i1}, ..., m_{in_i}$ be the possible moves in state s_i and m_{ib} be the move of the best sequence in state s_i. The goal is to learn to play the move m_{ib} in state s_i.

The playouts use Gibbs sampling. Each move m_{ik} is associated to a weight w_{ik}. The probability p_{ik} of choosing the move m_{ik} in a playout is the softmax function:

$$p_{ik} = \frac{e^{w_{ik}}}{\Sigma_j e^{w_{ij}}}$$

The cross-entropy loss for learning to play move m_{ib} is $C_i = -log(p_{ib})$. In order to apply the gradient we calculate the partial derivative of the loss: $\frac{\delta C_i}{\delta p_{ib}} = -\frac{1}{p_{ib}}$. We then calculate the partial derivative of the softmax with respect to the weights:

$$\frac{\delta p_{ib}}{\delta w_{ij}} = p_{ib}(\delta_{bj} - p_{ij})$$

Where $\delta_{bj} = 1$ if $b = j$ and 0 otherwise. Thus the gradient is:

$$\nabla w_{ij} = \frac{\delta C_i}{\delta p_{ib}} \frac{\delta p_{ib}}{\delta w_{ij}} = -\frac{1}{p_{ib}} p_{ib}(\delta_{bj} - p_{ij}) = p_{ij} - \delta_{bj}$$

If we use α as a learning rate we update the weights with:

$$w_{ij} = w_{ij} - \alpha(p_{ij} - \delta_{bj})$$

This is the formula used in the NRPA algorithm to adapt weights.

4   Generalization of NRPA

We propose to generalize the NRPA algorithm by generalizing the way the probability is calculated using a temperature τ and a bias β_{ij}:

$$p_{ik} = \frac{e^{\frac{w_{ik}}{\tau} + \beta_{ik}}}{\Sigma_j e^{\frac{w_{ij}}{\tau} + \beta_{ij}}}$$

4.1   Theoretical Analysis

The formula for the derivative of $f(x) = \frac{g(x)}{h(x)}$ is:

$$f'(x) = \frac{g'(x)h(x) - h'(x)g(x)}{h^2(x)}$$

So the derivative of p_{ib} relative to w_{ib} is:

$$\frac{\delta p_{ib}}{\delta w_{ib}} = \frac{\frac{1}{\tau} e^{\frac{w_{ib}}{\tau} + \beta_{ib}} \Sigma_j e^{\frac{w_{ij}}{\tau} + \beta_{ij}} - \frac{1}{\tau} e^{\frac{w_{ib}}{\tau} + \beta_{ib}} e^{\frac{w_{ib}}{\tau} + \beta_{ib}}}{(\Sigma_j e^{\frac{w_{ij}}{\tau} + \beta_{ij}})^2}$$

$$\frac{\delta p_{ib}}{\delta w_{ib}} = \frac{1}{\tau} \frac{e^{\frac{w_{ib}}{\tau} + \beta_{ib}}}{\Sigma_j e^{\frac{w_{ij}}{\tau} + \beta_{ij}}} \frac{\Sigma_j e^{\frac{w_{ij}}{\tau} + \beta_{ij}} - e^{\frac{w_{ib}}{\tau} + \beta_{ib}}}{\Sigma_j e^{\frac{w_{ij}}{\tau} + \beta_{ij}}}$$

$$\frac{\delta p_{ib}}{\delta w_{ib}} = \frac{1}{\tau} p_{ib}(1 - p_{ib})$$

The derivative of p_{ib} relative to w_{ij} with $j \neq b$ is:

$$\frac{\delta p_{ib}}{\delta w_{ij}} = -\frac{1}{\tau} \frac{e^{\frac{w_{ij}}{\tau} + \beta_{ij}} e^{\frac{w_{ib}}{\tau} + \beta_{ib}}}{(\Sigma_j e^{\frac{w_{ij}}{\tau} + \beta_{ij}})^2}$$

$$\frac{\delta p_{ib}}{\delta w_{ij}} = -\frac{1}{\tau} p_{ij} p_{ib}$$

We then derive the cross-entropy loss and the softmax to calculate the gradient:

$$\nabla w_{ij} = \frac{\delta C_i}{\delta p_{ib}} \frac{\delta p_{ib}}{\delta w_{ij}} = -\frac{1}{\tau} \frac{1}{p_{ib}} p_{ib}(\delta_{bj} - p_{ij}) = \frac{p_{ij} - \delta_{bj}}{\tau}$$

If we use α as a learning rate we update the weights with:

$$w_{ij} = w_{ij} - \alpha \frac{p_{ij} - \delta_{bj}}{\tau}$$

This is a generalization of NRPA since when we set $\tau = 1$ and $\beta_{ij} = 0$ we get NRPA.

The corresponding algorithms are given in Algorithms 4 and 5.

4.2   Equivalence of Algorithms

Let the weights and probabilities of playing moves be indexed by the iteration of the GNRPA level. Let w_{nij} be the weight w_{ij} at iteration n, p_{nij} be the probability of playing move j at step i at iteration n, δ_{nbj} the δ_{bj} at iteration n. We have:

$$p_{0ij} = \frac{e^{\frac{1}{\tau} w_{0ij} + \beta_{ij}}}{\Sigma_k e^{\frac{1}{\tau} w_{0ik} + \beta_{ik}}}$$

$$w_{1ij} = w_{0ij} - \frac{\alpha}{\tau}(p_{0ij} - \delta_{0bj})$$

$$p_{1ij} = \frac{e^{\frac{1}{\tau} w_{1ij} + \beta_{ij}}}{\Sigma_k e^{\frac{1}{\tau} w_{1ik} + \beta_{ik}}} = \frac{e^{\frac{1}{\tau} w_{0ij} - \frac{\alpha}{\tau^2}(p_{0ij} - \delta_{0bj}) + \beta_{ij}}}{\Sigma_k e^{\frac{1}{\tau} w_{1ik} + \beta_{ik}}}$$

$$w_{2ij} = w_{1ij} - \frac{\alpha}{\tau}(p_{1ij} - \delta_{1bj}) = w_{0ij} - \frac{\alpha}{\tau}(p_{0ij} - \delta_{0bj} + p_{1ij} - \delta_{1bj})$$

By recurrence we get:

$$p_{nij} = \frac{e^{\frac{1}{\tau} w_{nij} + \beta_{ij}}}{\Sigma_k e^{\frac{1}{\tau} w_{nik} + \beta_{ik}}} = \frac{e^{\frac{w_{0ij}}{\tau} - \frac{\alpha}{\tau^2}(\Sigma_k p_{kij} - \delta_{kbj}) + \beta_{ij}}}{\Sigma_k e^{\frac{1}{\tau} w_{nik} + \beta_{ik}}}$$

From this equation we can deduce the equivalence between different algorithms. For example GNRPA$_1$ with $\alpha_1 = (\frac{\tau_1}{\tau_2})^2 \alpha_2$ and τ_1 is equivalent to GNRPA$_2$ with α_2 and τ_2 provided we set w_{0ij} in GNRPA$_1$ to $\frac{\tau_1}{\tau_2} w_{0ij}$. It means we can always use $\tau = 1$ provided we correspondingly set α and w_{0ij}.

Another deduction we can make is we can set $\beta_{ij} = 0$ provided we set $w_{0ij} = w_{0ij} + \tau \times \beta_{ij}$. We can also set $w_{0ij} = 0$ and use only β_{ij} which is easier.

The equivalences mean that GNRPA is equivalent to NRPA with the appropriate α and w_{0ij}. However, it can be more convenient to use β_{ij} than to initialize the weights w_{0ij} as we will see for SameGame.

Algorithm 4. The generalized playout algorithm

```
1: playout (state, policy)
2:    sequence ← []
3:    while true do
4:       if state is terminal then
5:          return (score (state), sequence)
6:       end if
7:       z ← 0
8:       for m ∈ possible moves for state do
9:          o[m] ← e^( policy[code(m)]/τ + β(m) )
10:         z ← z + o[m]
11:      end for
12:      choose a move with probability  o[move]/z
13:      state ← play(state, move)
14:      sequence ← sequence + move
15:   end while
```

Algorithm 5. The generalized adapt algorithm

```
1: Adapt (policy, sequence)
2:    polp ← policy
3:    state ← root
4:    for b ∈ sequence do
5:       z ← 0
6:       for m ∈ possible moves for state do
7:          o[m] ← e^( policy[code(m)]/τ + β(m) )
8:          z ← z + o[m]
9:       end for
10:      for m ∈ possible moves for state do
11:         polp[code(m)] ← polp[code(m)] − α/τ ( o[m]/z − δ_bm )
12:      end for
13:      state ← play(state, b)
14:   end for
15:   policy ← polp
```

---

5   Optimizations of GNRPA

5.1   Avoid Calculating Again the Possible Moves

In problems such as SameGame the computation of the possible moves is costly. It is important in this case to avoid to compute again the possible moves for the best playout in the Adapt function. The possible moves have already been calculated during the playout that found the best sequence. The optimized playout algorithm memorizes in a matrix *code* the codes of the possible moves during a playout. The cell $code[i][m]$ contains the code of the possible move of index m at the state number i of the best sequence. The state number 0 is the initial state of the problem. The *index* array memorizes the index of the code of the best move for each state number, $len(index)$ is the length of the best sequence and $index[i]$ is the index of the best move for state number i.

5.2   Avoid the Copy of the Policy

Tha Adapt algorithm of NRPA and GNRPA considers the states of the sequence to learn as a batch. The sum of the gradients is calculated for the entire sequence and then applied. The way it is done in NRPA is by copying the policy to a temporary policy, modifying the temporary policy computing the gradient with the unmodified policy, and then copying the modified temporary policy to the policy.

When the number of possible codes is large copying the policy can be costly. We propose to change the Adapt algorithm to avoid to copy twice the policy at each Adapt call. We also use the memorized codes and index so as to avoid calculating again the possible moves of the best sequence.

The way to avoid copying the policy is to make a first loop to compute the probabilities of each move of the best sequence, lines 2–8 of Algorithm 6. The matrix $o[i][m]$ contains the probability for move index m in state number i, the array $z[i]$ contains the sum of the probabilities of state number i. The second step is to apply the gradient directly to the policy for each state number i and each code, see lines 9–14.

Algorithm 6. The optimized generalized adapt algorithm

1: Adapt $(policy, code, index)$
2:　　**for** $i \in [0, len(index)[$ **do**
3:　　　$z[i] \leftarrow 0$
4:　　　**for** $m \in [0, len(code[i])[$ **do**
5:　　　　$o[i][m] \leftarrow e^{\frac{policy[code[i][m]]}{\tau} + \beta(m)}$
6:　　　　$z[i] \leftarrow z[i] + o[i][m]$
7:　　　**end for**
8:　　**end for**
9:　　**for** $i \in [0, len(index)[$ **do**
10:　　　$b \leftarrow index[i]$
11:　　　**for** $m \in [0, len(code[i])[$ **do**
12:　　　　$policy[code[i][m]] \leftarrow policy[code[i][m]] - \frac{\alpha}{\tau}(\frac{o[i][m]}{z[i]} - \delta_{bm})$
13:　　　**end for**
14:　　**end for**

6　Experimental Results

We now give experimental results for SameGame and TSPTW.

6.1　SameGame

The first algorithm we test is the standard NRPA algorithm with codes of the moves using a Zobrist hashing [24] of the cells of the moves [6,10,17]. The selective policy used is to avoid the moves of the dominant color except for moves of size two after move number ten. The codes of the possible moves of the best playout are recorded so as to avoid computing again the possible moves in the Adapt function. It is called NRPA.

Using Zobrist hashing of the moves and biasing the policy with β is better than initializing the weights at SameGame since there are too many possible moves and weights. We tried to reduce the possible codes for the moves but it gave worse results. The second algorithm we test is to use Zobrist hashing and the selective policy associated to the bias. It is GNRPA with $\tau = 1$ and $\beta_{ij} = min(n - 2 - tabu, 8)$, with $tabu = 1$ if the move is of size 2 and of the tabu color and $tabu = 0$ otherwise. The variable n being the number of cells of the move. The algorithm is called GNRPA.beta.

The third algorithm we test is to use Zobrist hashing, the selective policy, β and the optimized Adapt function. The algorithm is called GNRPA.beta.opt.

All algorithms are run 200 times for 655.36 s and average scores are recorded each time the search time is doubled.

The evolution of the average score of the algorithms is given in Fig. 1. We can see that GNRPA.beta is better than NRPA but that for scores close to the current record of the problem the difference is small. GNRPA.beta.opt is the best algorithm as it searches more than GNRPA.beta for the same time.

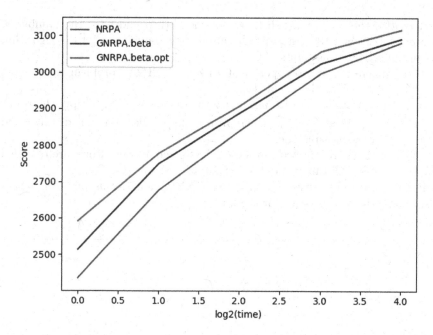

Fig. 1. Evolution of the average scores of the three algorithms at SameGame.

Table 1 gives the average scores for the three algorithms associated to the 95% confidence interval in parenthesis ($2 \times \frac{\sigma}{\sqrt{n}}$).

Table 1. Results for the first SameGame problem of the standard test suite.

Time	NRPA	GNRPA.beta	GNRPA.beta.opt
40.96	2435.12 (49.26)	2513.35 (53.57)	2591.46 (52.50)
81.92	2676.39 (47.16)	2749.33 (47.82)	2777.83 (48.05)
163.84	2838.99 (41.82)	2887.78 (39.50)	2907.23 (38.45)
327.68	2997.74 (21.39)	3024.68 (18.27)	3057.78 (13.52)
655.36	3081.25 (10.66)	3091.44 (10.96)	3116.54 (7.42)

6.2 TSPTW

The Traveling Salesman with Time Windows problem (TSPTW) is a practical problem that has everyday applications. NRPA can be used to efficiently solve practical logistics problems faced by large companies such as EDF [7].

In NRPA paths with violated constraints can be generated. As presented in [22], a new score $Tcost(p)$ of a path p can be defined as follow:

$$Tcost(p) = cost(p) + 10^6 * \Omega(p),$$

with, $cost(p)$ the sum of the distances of the path p and $\Omega(p)$ the number of violated constraints. 10^6 is a constant chosen high enough so that the algorithm first optimizes the constraints.

The problem we use to experiment with the TSPTW problem is the most difficult problem from the set of [19].

In order to initialize β_{ij} we normalize the distances and multiply the result by ten. So $\beta_{ij} = 10 \times \frac{d_{ij}-min}{max-min}$, where min is the smallest possible distance and max the greatest possible one.

All algorithms are run 200 times for 655.36 s and average scores are recorded each time the search time is doubled.

Figure 2 gives the curves for the three GNRPA algorithms we haves tested with a logarithmic time scale for the x axis.

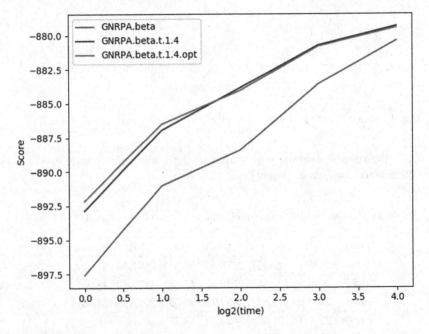

Fig. 2. Evolution of the average scores of the three algorithms for TSPTW.

We could not represent the curve for NRPA in Fig. 2 since the average values are too low. They are given in Table 2. It is possible to improve much on standard NRPA by initializing the weights with the distances between cities [7,11]. However this solution is not practical for all problems as we have seen with SameGame and using a bias β is more convenient and general. We also tried initializing the weights with β instead of using β and we got similar results to the use of β.

We can see in Fig. 2 that using a temperature of 1.4 improves on a temperature of 1.0. Using the optimized Adapt function does not improve GNRPA for TSPTW since in the TSPTW problem the policy array and the number of possible moves is very small and copying the policy is fast.

The curve of the best algorithm is asymptotic toward the best value found by all algorithms. It reaches better scores faster.

Table 2 gives the average values for NRPA and the three GNRPA algorithms we have tested. As there is a penalty of 1 million for each constraint violation, NRPA has very low scores compared to GNRPA. This is why NRPA is not depicted in Fig. 2. For a search time of 655.36 s and not taking into account the constraints, NRPA usually reaches tour scores between −900 and −930. Much worse than GNRPA. We can observe that using a temperature is beneficial until we use 655.36 s and approach the asymptotic score when both algorithms have similar scores. The numbers in parenthesis in the table are the 95% confidence interval ($2 \times \frac{\sigma}{\sqrt{n}}$).

Table 2. Results for the TSPTW rc204.1 problem

Time	NRPA	GNRPA.beta	GNRPA.beta.t.1.4	GNRPA.beta.t.1.4.opt
40.96	−3745986.46 (245766.53)	−897.60 (1.32)	−892.89 (0.96)	−892.17 (1.04)
81.92	−1750959.11 (243210.68)	−891.04 (1.05)	−886.97 (0.87)	−886.52 (0.83)
163.84	−1030946.86 (212092.35)	−888.44 (0.98)	−883.87 (0.71)	−884.07 (0.70) .
327.68	−285933.63 (108975.99)	−883.61 (0.63)	−880.76 (0.40)	−880.83 (0.32)
655.36	−45918.97 (38203.97)	−880.42 (0.30)	−879.35 (0.16)	−879.45 (0.17)

7   Conclusion

We presented a theoretical analysis and a generalization of NRPA named GNRPA. It uses a temperature τ and a bias β.

We have theoretically shown that using a bias is equivalent to initializing the weights. For SameGame initializing the weights can be difficult if we initialize all the weights at the start of the program since there are too many possible weights, whereas using a bias β is easier and improves search at SameGame. A lazy initialization of the weights would also be possible in this case and would solve the weight initialization problem for SameGame. For some other problems the bias could be more specific than the code of the move, i.e. a move could be associated to different bias depending on the state. In this case different bias could be used in different states for the same move which would not be possible with weight initialization.

We have also theoretically shown that the learning rate and the temperature can replace each other. Tuning the temperature and using a bias has been very beneficial for the TSPTW.

The remaining work is to apply the algorithm to other domains and to improve the way to design formulas for the bias β.

Acknowledgment. This work was supported in part by the French government under management of Agence Nationale de la Recherche as part of the "Investissements d'avenir" program, reference ANR19-P3IA-0001 (PRAIRIE 3IA Institute).

References

1. Bjornsson, Y., Finnsson, H.: CadiaPlayer: a simulation-based general game player. IEEE Trans. Comput. Intell. AI Games **1**(1), 4–15 (2009)
2. Bouzy, B.: Monte-Carlo fork search for cooperative path-finding. In: Cazenave, T., Winands, M.H.M., Iida, H. (eds.) CGW 2013. CCIS, vol. 408, pp. 1–15. Springer, Cham (2014). https://doi.org/10.1007/978-3-319-05428-5_1
3. Bouzy, B.: Burnt pancake problem: new lower bounds on the diameter and new experimental optimality ratios. In: Proceedings of the Ninth Annual Symposium on Combinatorial Search, SOCS 2016, Tarrytown, NY, USA, 6–8 July 2016, pp. 119–120 (2016)
4. Browne, C., et al.: A survey of Monte Carlo tree search methods. IEEE Trans. Comput. Intell. AI Games **4**(1), 1–43 (2012). https://doi.org/10.1109/TCIAIG.2012.2186810
5. Cazenave, T.: Nested Monte-Carlo search. In: Boutilier, C. (ed.) IJCAI, pp. 456–461 (2009)
6. Cazenave, T.: Nested rollout policy adaptation with selective policies. In: Cazenave, T., Winands, M.H.M., Edelkamp, S., Schiffel, S., Thielscher, M., Togelius, J. (eds.) CGW/GIGA 2016. CCIS, vol. 705, pp. 44–56. Springer, Cham (2017). https://doi.org/10.1007/978-3-319-57969-6_4
7. Cazenave, T., Lucas, J.Y., Kim, H., Triboulet, T.: Monte Carlo vehicle routing (2020, submitted)
8. Cazenave, T., Saffidine, A., Schofield, M.J., Thielscher, M.: Nested Monte Carlo search for two-player games. In: Proceedings of the Thirtieth AAAI Conference on Artificial Intelligence, Phoenix, Arizona, USA, 12–17 February 2016, pp. 687–693 (2016). http://www.aaai.org/ocs/index.php/AAAI/AAAI16/paper/view/12134
9. Cazenave, T., Teytaud, F.: Application of the nested rollout policy adaptation algorithm to the traveling salesman problem with time windows. In: Hamadi, Y., Schoenauer, M. (eds.) LION 2012. LNCS, pp. 42–54. Springer, Heidelberg (2012). https://doi.org/10.1007/978-3-642-34413-8_4
10. Edelkamp, S., Cazenave, T.: Improved diversity in nested rollout policy adaptation. In: Friedrich, G., Helmert, M., Wotawa, F. (eds.) KI 2016. LNCS (LNAI), vol. 9904, pp. 43–55. Springer, Cham (2016). https://doi.org/10.1007/978-3-319-46073-4_4
11. Edelkamp, S., Gath, M., Cazenave, T., Teytaud, F.: Algorithm and knowledge engineering for the TSPTW problem. In: 2013 IEEE Symposium on Computational Intelligence in Scheduling (SCIS), pp. 44–51. IEEE (2013)
12. Edelkamp, S., Gath, M., Greulich, C., Humann, M., Herzog, O., Lawo, M.: Monte-Carlo tree search for logistics. In: Clausen, U., Friedrich, H., Thaller, C., Geiger, C. (eds.) Commercial Transport. LNL, pp. 427–440. Springer, Cham (2016). https://doi.org/10.1007/978-3-319-21266-1_28
13. Edelkamp, S., Gath, M., Rohde, M.: Monte-Carlo tree search for 3D packing with object orientation. In: Lutz, C., Thielscher, M. (eds.) KI 2014. LNCS (LNAI), vol. 8736, pp. 285–296. Springer, Cham (2014). https://doi.org/10.1007/978-3-319-11206-0_28

14. Edelkamp, S., Greulich, C.: Solving physical traveling salesman problems with policy adaptation. In: 2014 IEEE Conference on Computational Intelligence and Games (CIG), pp. 1–8. IEEE (2014)

15. Edelkamp, S., Tang, Z.: Monte-Carlo tree search for the multiple sequence alignment problem. In: Eighth Annual Symposium on Combinatorial Search (2015)

16. Méhat, J., Cazenave, T.: Combining UCT and nested Monte Carlo search for single-player general game playing. IEEE Trans. Comput. Intell. AI Games $2(4)$, 271–277 (2010)

17. Negrevergne, B., Cazenave, T.: Distributed nested rollout policy for SameGame. In: Cazenave, T., Winands, M.H.M., Saffidine, A. (eds.) CGW 2017. CCIS, vol. 818, pp. 108–120. Springer, Cham (2018). https://doi.org/10.1007/978-3-319-75931-9_8

18. Portela, F.: An unexpectedly effective Monte Carlo technique for the RNA inverse folding problem. bioRxiv, p. 345587 (2018)

19. Potvin, J.Y., Bengio, S.: The vehicle routing problem with time windows part II: genetic search. INFORMS J. Comput. $8(2)$, 165–172 (1996)

20. Poulding, S.M., Feldt, R.: Generating structured test data with specific properties using nested Monte-Carlo search. In: Genetic and Evolutionary Computation Conference, GECCO 2014, Vancouver, BC, Canada, 12–16 July 2014, pp. 1279–1286 (2014)

21. Poulding, S.M., Feldt, R.: Heuristic model checking using a Monte-Carlo tree search algorithm. In: Proceedings of the Genetic and Evolutionary Computation Conference, GECCO 2015, Madrid, Spain, 11–15 July 2015, pp. 1359–1366 (2015)

22. Rimmel, A., Teytaud, F., Cazenave, T.: Optimization of the nested Monte-Carlo algorithm on the traveling salesman problem with time windows. In: Di Chio, C., et al. (eds.) EvoApplications 2011. LNCS, vol. 6625, pp. 501–510. Springer, Heidelberg (2011). https://doi.org/10.1007/978-3-642-20520-0_51

23. Rosin, C.D.: Nested rollout policy adaptation for Monte Carlo Tree Search. In: IJCAI, pp. 649–654 (2011)

24. Zobrist, A.L.: A new hashing method with application for game playing. ICCA J. $13(2)$, 69–73 (1970)

Monte Carlo Inverse Folding

Tristan Cazenave[✉] and Thomas Fournier

LAMSADE, Université Paris-Dauphine, PSL, CNRS, Paris, France
Tristan.Cazenave@dauphine.psl.eu, thomas.fournier@dauphine.eu

Abstract. The RNA Inverse Folding problem comes from computational biology. The goal is to find a molecule that has a given folding. It is important for scientific fields such as bioengineering, pharmaceutical research, biochemistry, synthetic biology and RNA nanostructures. Nested Monte Carlo Search has given excellent results for this problem. We propose to adapt and evaluate different Monte Carlo Search algorithms for the RNA Inverse Folding problem.

1 Introduction

Monte Carlo Tree Search (MCTS) [17,25] originated in the field of computer Go and has been applied to many other problems [4].

Nested Monte Carlo Search (NMCS) [5] uses multiple levels of search, memorizing the best sequence of each level. It has been applied to many single-player games and optimization problems [2,3,7,24,26,30–32] and also to two-player games [8].

Nested Rollout Policy Adaptation (NRPA) [33] also uses multiple levels of search, memorizing the best sequences. It additionally learns a playout policy using the best sequences. It has also been applied to many problems [12,14,18–23] and games [9].

The RNA design problem also named the RNA Inverse Folding problem is computationally hard [1]. This problem is important for scientific fields such as bioengineering, pharmaceutical research, biochemistry, synthetic biology and RNA nanostructures [29]. NMCS has been successfully applied to the RNA Inverse Folding problem with the NEMO program [29]. As a follow-up to NEMO, we propose to investigate different Monte Carlo Search algorithms for this problem.

The paper is organized as follows. The second section describes the Inverse Folding problem, the NEMO program by Fernando Portela [29] and the domain knowledge used in NEMO. The third section describes the Monte Carlo Search algorithms we have used for solving Inverse Folding problems of the Eterna100 benchmark. We present a new algorithm performing well for this problem, the Generalized Nested Rollout Policy Adaptation (GNRPA) algorithm with restarts. The fourth sections details experimental results.

© Springer Nature Switzerland AG 2021
T. Cazenave et al. (Eds.): MCS 2020, CCIS 1379, pp. 84–99, 2021.
https://doi.org/10.1007/978-3-030-89453-5_7

2   Inverse Folding

2.1   Presentation of the RNA Inverse Folding Problem

An RNA strand is a molecule composed of a sequence of nucleotides. This strand folds back on itself to form what is called its secondary structure (See Fig. 1). It is possible to find in a polynomial time the folded structure of a given sequence. However, the opposite, which is the RNA inverse folding problem, is much harder and is supposed to be NP-complete. This problem still resists algorithmic approaches that still fail to match the performance of human experts. This is partly due to the chaotic changes that can occur in the secondary structure because of a small change in the sequence (See difference between Fig. 3 and 2b).

These performances are evaluated on the Eterna100 benchmark which contains 100 RNA secondary structure puzzles of varying degrees of difficulty. A puzzle consists of a given structure under the dot-bracket notation. This notation defines a structure as a sequence of parentheses and points each representing a base. The matching parentheses symbolize the paired bases and the dots the unpaired ones. The puzzle is solved when a sequence of the four nucleotides A, U, G and C, folding according to the target structure, is found. In some puzzles, the value of certain bases is imposed.

Where human experts have managed to solve the 100 problems of the benchmark, no program has so far achieved such a score. The best score so far is 95/100 by NEMO, NEsted MOnte Carlo RNA Puzzle Solver [29].

2.2   NEMO

NEMO works by performing several iterations of NMCS-B, a slightly modified version of NMCS. Between each iteration of the NMCS-B, NEMO retains part of the best current solution. It identifies a part of the sequence on which to perform mutations using stochastic heuristics and restarts the NMCS-B on it. We model a candidate by a sequence. The bases assigned are represented by the corresponding letter (A/U/G/C) and the others by the letter N. This is how the NMCS-B identifies the bases on which it must work.

NEMO uses a level 1 NMCS for its NMCS-B. At level 1, the NMCS, in each state of the problem, will perform a certain number of playouts for each possible move. It then plays the move that led to the best playout and moves to the next state until it reaches a final state. In the context of NEMO, each state corresponds to a sequence, initially the candidate sequence. Each move consists in taking the first N in the sequence and assigning a value to it, working on paired bases first. When the base is paired in the target structure, it will assign a value to both bases of the pair simultaneously. Indeed only the three combinations AU, GC, and GU can be paired, so it is more convenient to consider them at the same time. To perform playouts, NEMO is also using heuristics with biased weights depending on the location in the target structure. In addition, unlike the classic NMCS, the NMCS-B retains the best playout achieved so far throughout

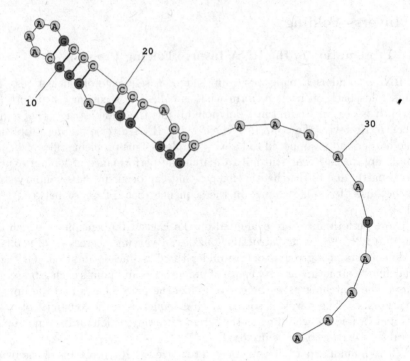

Fig. 1. Secondary structure of a RNA strand

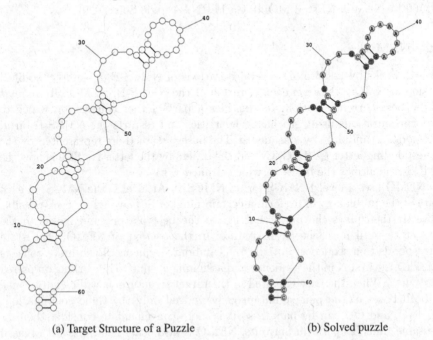

(a) Target Structure of a Puzzle   (b) Solved puzzle

Fig. 2. Exemple of an Eterna100 puzzle

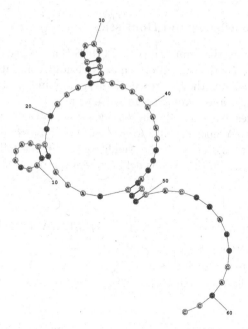

Fig. 3. Sequence of Fig. 2b with base 25 replaced by U

the execution. A final state is found when the sequence is fully completed. The playouts are evaluated according to the function:

$$score = \begin{cases} \frac{K}{1+\Delta G} & \text{if } K > 0 \\ K(1 + \Delta G) & \text{else} \end{cases}$$

$$\text{with } K = 1 - \frac{BPD}{2 * NumTargetPairs}$$

Where BPD is the number of different pairs between the secondary structure of the sequence and the target structure.
$NumTargetPairs$ is the number of pairs in the target structure.
ΔG is the difference between the Minimum Free Energy of the secondary structure and the free energy that the sequence would have in the target structure.

The objective is to maximize the score function until a value of 1 is obtained, meaning that the problem is solved. Indeed, we use a model based on Minimum Free Energy (MFE) predictions. For each pair composed of a sequence and a secondary structure of the same size we can compute the Gibbs-Free Energy. The predicted secondary structure of a sequence is the one minimizing the Gibbs-Free Energy among all structures of this length. Thus we try to minimize the positive values ΔG and BPD until they are zero, i.e. score $= 1$. Due to the large number of possible matches, BPD is for most sequences very high especially for long sequences and K is therefore negative. However, interesting sequences whose secondary structure is close to the target structure thus have scores between 0 and 1.

2.3   Domain Knowledge and Heuristics

The heuristics used for the sampling of the NMCS-B are based on domain knowledge and personal experience and are chosen without computational optimization. These are crucial as they increase the average number of puzzles solved in a run from 88 to 92.1 and is on average twice as fast.

The paired bases are generally chosen according to the same rule. With the exception of adjacent stacks in multi-loops, the closing pairs of the left-most and rightmost stacks are chosen with the weights given in the following table. The notion of right and left in this case is defined from the point of view of the inside of the loop.

Paired Bases	GC/CG	AU/UA	GU/UG
General Case	60%	33%	7%
Left-Most in Junction	82%	11%	7%
Right-Most in Junction	37%	56%	7%

Various rules are applied for unpaired bases. The weights used to choose between A/U/G/C in the general case are 93%, 1%, 5% and 1%. Mismatches are treated differently depending on the case.

Since NEMO first assigns a value to the paired bases, the weights for the bases with a paired mismatch depend on the value of this mismatch.

Mismatch with a Paired Base	A	U	G	C
Mismatch is a paired A	63%	0%	25%	12%
Mismatch is a paired U	0%	55%	9%	36%
Mismatch is a paired G	25%	12%	63%	0%
Mismatch is a paired C	55%	36%	0%	9%

Furthermore, in internal loops, the weights also depend on the mismatch value if it has already been assigned, otherwise a more general rule applies.

Mismatch in Internal Loops	A	U	G	C
Mismatch is not assigned	18%	4%	74%	4%
Mismatch is A	44%	0%	44%	12%
Mismatch is U	0%	67%	11%	22%
Mismatch is G	67%	11%	22%	0%
Mismatch is C	66%	17%	0%	17%

Finally, the mismatch in junctions and external loops are drawn according to the distribution 97%, 1%, 1% and 1%.

Much stronger and more deterministic rules are applied with high probabilities (more than 80%) in specific cases, especially for triloops and internal loops. For both 1-1 and 2-2 internal loops for instance, only one mismatch pair is possible, and there are only three possibilities in the general case. This is part of a process of reproducing a "boosting" strategy. Depending on the type of loop, certain combinations of nucleotides at specific locations called "boosting points", especially terminal mismatches, can be used to reduce the energy of the structure. However, the most difficult puzzles may require less conventional solutions, hence the need not to apply these rules 100% of the time.

Therefore, in the use we will make of this heuristic we will not apply these last rules and when we mention the weights of the NEMO heuristic we refer to the previously mentioned values.

In addition, between iterations of NMCS-B, if it hasn't solved the problem NEMO keeps part of the best current solution to restart the algorithm on. The set of bases that are not kept contains those that do not fold correctly, their neighborhood and randomly selected bases. This principle, i.e. keeping part of the best current solution, has not been applied to the presented algorithms.

3 Monte Carlo Search

3.1 Presentation of the NRPA Algorithm

The Nested Rollout Policy Adaptation (NRPA) algorithm is a Monte Carlo Tree Search based algorithm with adaptive rollout policy during execution. It is a recursive algorithm. At level 0 it generates a playout according to the current policy. At level n, it calls for a given number of iterations the n-1 level of the algorithm, adapting the policy each time with the best solution so far. NRPA is given in Algorithms 1, 2 and 3. The policy consists of a set of weights, one for each possible coding value. The adaptation is made through gradient ascent. Given a rollout to adapt the policy on, we increase the weights of the moves chosen during this playout in proportion to their values and decreasing that of the other possible moves in each state. The NRPA applies this gradient ascent on the best current solution of the level after each iteration.

Algorithm 1. The playout algorithm

```
 1: playout (state, policy)
 2:    sequence ← []
 3:    while true do
 4:       if state is terminal then
 5:          return (score (state), sequence)
 6:       end if
 7:       z ← 0.0
 8:       for m in possible moves for state do
 9:          z ← z + exp (policy [code(m)])
10:       end for
11:       choose a move with probability exp(policy[code(move)])/z
12:       state ← play (state, move)
13:       sequence ← sequence + move
14:    end while
```

Algorithm 2. The Adapt algorithm

```
 1: Adapt (policy, sequence)
 2:    polp ← policy
 3:    state ← root
 4:    for move in sequence do
 5:       polp [code(move)] ← polp [code(move)] + α
 6:       z ← 0.0
 7:       for m in possible moves for state do
 8:          z ← z + exp (policy [code(m)])
 9:       end for
10:       for m in possible moves for state do
11:          polp [code(m)] ← polp [code(m)] - α * exp(policy[code(m)])/z
12:       end for
13:       state ← play (state, move)
14:    end for
15:    policy ← polp
```

Algorithm 3. The NRPA algorithm.

```
 1: NRPA (level, policy)
 2:    if level == 0 then
 3:       return playout (root, policy)
 4:    else
 5:       bestScore ← −∞
 6:       for N iterations do
 7:          (result,new) ← NRPA(level − 1, policy)
 8:          if result ≥ bestScore then
 9:             bestScore ← result
10:             seq ← new
11:          end if
12:          policy ← Adapt (policy, seq)
13:       end for
14:       return (bestScore, seq)
15:    end if
```

3.2   Application of NRPA to Inverse Folding

As part of the Inverse Folding problem, one solution consists in a chain of bases. A playout is made by running through the targeted structure of the chain, each move consists in assigning a value to the missing links. We distinguish between two cases, the unpaired bases in the target structure have four possible moves, one for each nucleic base (A/U/G/C). The paired bases have 6 possible moves, one for each possible ordered combination with their pair (GC/CG/AU/...). Each move is therefore defined by its position in the chain and whether it is a pair or not. Thus, there is a fixed number of moves, which are always ordered in the same way. Solutions are evaluated with the same score function as in the NEMO algorithm which is a combination of the fitness of the chain with the target structure and the difference between the target structure and the folded chain structure.

3.3   Generalized NRPA

Let w_{ib} be the weight associated to move b at index i in the sequence. In NRPA the probability of choosing move b at index i is:

$$p_{ib} = \frac{e^{w_{ib}}}{\Sigma_k e^{w_{ik}}}$$

We propose to try Generalized NRPA (GNRPA) [10] for Inverse Folding and to replace it with:

$$p_{ib} = \frac{e^{w_{ib}+\beta_{ib}}}{\Sigma_k e^{w_{ik}+\beta_{ik}}}$$

where we use for β_{ib} the logarithm of the probabilities used in NEMO.

3.4   Stabilized GNRPA

Stabilized NRPA [13] is a simple improvement of NRPA. The principle is to play P playouts at level 1 before each call to the adapt function. The number of calls to the adapt function as level 1 is still N, the number of iteration of upper levels. So at level 1, $P \times N$ playouts are performed. The same principle is used for stabilized GNRPA.

3.5   Beam GNRPA

Monte Carlo Beam Search [6] and Beam NRPA [15] have already been applied successfully to the TSPTW and to Morpion Solitaire. The best results were obtained using a beam at level 1. Similarly to Stabilized GNRPA, at level 1, $B \times N$ playouts are performed for a beam of size B. However the algorithm is different from Stabilized GNRPA since it memorizes B best sequences and B policies and plays the B playouts with different policies.

As Stabilized GNRPA it is embarrassingly parallel at level 1 and can be very efficient on a parallel machine.

When using Beam NRPA/GNRPA it can be beneficial to ensure the diversity of the beam [18]. The diversity criterion we have used is to only keep in the beam sequences that have different scores. It is simple and efficient as it ensures diversity while keeping enough sequences.

3.6 Coding Moves

Moves in the policy are represented by their encoding values, hence their importance. The natural way to code moves for Inverse Folding is to use the index of the base or of the pair of base in the string (m.index) and the index of the base in the list of bases or of the pair of bases in the list of pairs of bases (m.number). The formula is then:

$$code(m) = m.index + M \times m.number$$

For example if the move m is to put the fourth base at index 10 the code is $10 + 2000 * 4 = 8010$, provided strings always have less than 2000 characters and therefore $M = 2000$.

It may be interesting to include the previously chosen bases in the code of a move, for example if a base has meaning only if following another base. We call the history of a code the number of bases in the history included in the code. The previous formula holds for a code history of 0. The code for a code history of 1 is:

$$code(m) = m.index + M \times m.number + M \times 6 \times previousMove.number$$

Six is the maximum value for a move number, the maximum number of legal moves is 6. The code for a code history of 2 includes the two previous moves in the code.

3.7 Start Learning

In order to wait for better sequences before learning it is possible to delay learning only after a given number of sequences have been found [8]. For this we call the adapt method only after a certain number of iterations as in line 19 of algorithm 4.

3.8   Zobrist Hashing

Each state is associated to a different hash code. This is done with Zobrist Hashing. Each move at each index is associated to a random number. The hash code of a state is the XOR of all the random numbers corresponding to the moves that have been played to reach this state. Zobrist hashing is used in games to build a transposition table. We will use it in the UCT variants. Another use of Zobrist Hashing is to detect playouts that have already been evaluated. As most of the time in Inverse Folding is spent scoring the playouts, it is advisable to avoid reevaluating an already evaluated playouts. This is done with a score hash table that contains the hash of the terminal states already encountered associated to their computed scores. If a terminal state is met again the score need not be recomputed it can just be sent back from the table. There are variations on the number of playouts already evaluated according to the sequence, some sequences have very few while others have a lot.

3.9   Restarts

There can be large variations on the solving times of some problems. Sometimes the search algorithm takes a wrong direction and stay stuck on a suboptimal sequence without making any progress. A way to deal with this behavior is to periodically stop and restart the search. For some difficult problems however a long search is required to find the solution. There are multiple ways to use restarts. The algorithm can double the search time at each restart for example. We call this method iterative doubling. We have observed that a level 2 search is able to solve many problems, a restart strategy can also be to repeatedly call GNRPA at level 2 until thinking time is elapsed. Another way to deal with search being stuck is to stop a level when the best sequence has not changed for a given number of recursive calls.

It is difficult to set a static restart strategy for all problems. Long sequences are much more difficult than short ones and the progress on long sequences is slower. In order to cope with this property we use a restart threshold. It is set to the length of the sequence divided by 5. When using this restart strategy there is no limit on the length of a level.

Algorithm 4 gives the GNRPA algorithm with restarts.

Algorithm 4. The GNRPA algorithm with restarts.

```
1: GNRPA (level, policy)
2:    if level == 0 then
3:        return playout (root, policy)
4:    else
5:        bestScore ← −∞
6:        last ← 0
7:        for i in range(∞) do
8:            (result,new) ← GNRPA(level − 1, policy)
9:            if result ≥ bestScore then
10:               if result > bestScore then
11:                   last ← i
12:               end if
13:               bestScore ← result
14:               seq ← new
15:           end if
16:           if i − last ≥ limitRestart then
17:               return (bestScore, seq)
18:           end if
19:           if i ≥ startLearning then
20:               policy ← Adapt (policy, seq)
21:           end if
22:       end for
23:   end if
```

3.10 Parallelization

The multiple playouts of stabilized NRPA and the loop over the elements of the beam are embarrassingly parallel. We simply parallelized with OpenMP a common loop including the beam and the stabilized playouts at level 1. If we have 4 stabilized playouts and a beam of 8, the 32 resulting playouts are played in parallel. This kind of parallelization is a kind of leaf parallelization [11,16].

We also experiment with root parallelization [11,16]. The principle is to perform multiple independent GNRPA in parallel and to stop as soon as one has found a solution or when the allocated time is elapsed.

Leaf parallelization is more difficult to scale than root parallelization. For the same wall clock time leaf parallelization runs more iterations for a single policy than root parallelization which optimizes many more different policies but with less iterations. On the other hand root parallelization scales very well and has a built-in restart strategy. Root parallelization works well for problems that converge relatively rapidly on a suboptimal solution as they benefit from restarts, while leaf parallelization works better for problems that converge slowly but steadily towards the best solutions.

Table 1. Number of problems solved, out of 100 problems, using different parameters.

Level	α	N	β_{ib}	P	Beam	H	Solved
1	1.0	100	no	1	1.1.1	0	3
1	1.0	100	yes	1	1.1.1	0	30
1	1.0	100	yes	1	1.1.1	1	32
1	1.0	100	yes	1	4.1.1	1	42
1	1.0	100	yes	1	8.1.1	1	53
1	1.0	100	yes	1	16.1.1	1	54
1	1.0	100	yes	4	8.1.1	1	69
1	1.0	100	yes	4	8.1.1	2	69
2	1.0	100	no	1	1.1.1	0	49
2	1.0	100	yes	1	1.1.1	0	73
2	1.0	100	yes	1	1.1.1	1	75
2	1.0	100	yes	2	1.1.1	0	73
2	1.0	100	yes	3	1.1.1	0	74
2	1.0	100	yes	4	1.1.1	0	80
2	1.0	100	yes	5	1.1.1	0	77
2	1.0	100	yes	6	1.1.1	0	75
2	1.0	100	yes	7	1.1.1	0	80
2	1.0	100	yes	8	1.1.1	0	79
2	1.0	100	yes	9	1.1.1	0	81
2	1.0	100	yes	10	1.1.1	0	80
2	1.0	100	yes	4	8.1.1	1	85
3	1.0	100	yes	1	1.1.1	0	85

Table 2. Parallelization efficiency.

Algorithm	1	2	4	6	8	12
GNRPA (level = 1, N = 100, P = 4, Beam = 8)	11.916	6.889	4.526	3.657	3.169	3.359

Table 3. Number of problems solved by GNRPA using different parameters and a fixed time limit.

β_{ib}	P	Beam	//	R	N	Start	H	1 m	2 m	4 m	8 m	16 m	32 m	64 m
no	1.1	1.1	n	n	100.100	0.0	0	30	33	41	50	61	67	69
yes	1.1	1.1	n	n	100.100	0.0	0	58	64	68	72	74	75	79
yes	4.1	4.1	n	n	100.100	0.0	0	71	75	78	79	81	83	84
yes	4.1	8.1	n	n	100.100	0.0	0	75	75	79	80	81	83	84
yes	4.1	8.1	n	n	100.100	0.0	1	75	78	80	82	83	85	87
yes	4.1	8.1	n	n	100.100	4.4	1	76	80	81	82	84	85	87
yes	4.1	8.1	y	n	100.100	4.4	1	78	84	83	86	87	87	88
yes	4.1	8.1	y	3	∞.∞	4.4	1	80	84	85	85	88	89	92

Table 4. Number of problems solved by different algorithms and a fixed time limit.

Algorithm	1 m	2 m	4 m	8 m	16 m	32 m	64 m
UCT(0.4)	47	49	50	54	56	57	62
Nested(1)	59	62	64	65	67	71	73
Nested(2)	55	59	61	66	66	71	73
Diversity GNRPA(5.1)	72	73	73	77	78	78	80

Table 5. Number of problems solved with different options.

Correction	Order	1 m	2 m	4 m	8 m	16 m
no	no	84	87	85	85	87
yes	no	80	86	85	86	87
no	yes	81	84	86	85	87
yes	yes	80	82	84	88	87

Table 6. Number of problems solved with root parallel GNRPA.

Process	1 m	2 m	4 m	8 m	16 m	32 m	64 m
20	82	84	85	86	87	88	89

4   Experimental Results

We now detail experiments with the different Monte Carlo Search algorithms on the Eterna100 benchmark.

Table 1 gives the number of problems solved with different parameters for the GNRPA algorithm. We can see that at level 1 using GNRPA instead of NRPA enables to solve 30 problems instead of 3. Similarly at level 2 it solves 73 problems instead of 49. Using Stabilized GNRPA with P = 4 and Beam GNRPA with a beam of 8 at level 1 also improves quite much the number of problems solved at levels 1 and 2. P and H take several values for the number of threads and beams, respectively, used at each level. Interestingly for level 2 it is 32 times slower that a regular level 2 search and solves 85 problems, when a search at level 3 is 100 times slower and still solves 85 problems.

Table 2 gives the median time over 3 runs of a level 1 search with different numbers of threads on problem 64 which is difficult. The algorithm is GNRPA with P = 4, a beam of 8 and N = 100. We can see that using 8 threads gives the best results. We optimized memory in order to avoid cache misses but we were not able to have better results with more threads.

Table 3 gives the number of problems solved within a fixed time limit for different algorithms. The time limits range from 1 min to 64 min per problem. The parallel algorithm runs on a multicore machine. The results of the parallel program are given in the last line of Table 3. The times used to stop the parallel program are the wall clock times. The last line is leaf parallel GNRPA with restarts and gives the best results within 1 h of wall clock time, solving 92 of the 100 problems in one run. Start is 4 meaning that it starts learning after 4 playouts. H is 1 meaning the code include the previous move, R is 3 meaning that the restart threshold is set to the length of the string divided by 3.

The problems solved by different runs of 1 h we made are not always the same. Some hard problems are solved only in some runs. So the limit of 92 solved problems is not the limit of the algorithm. The problems that were never solved in 1 h are problems 100, 99, 97, 91, 90, 78. With a two hours limit, problem 90 is solved thus reaching 95 solved problems, the same number of solved problems as NEMO. However, it is interesting to note that GNRPA solves problem 94 while NEMO solves problem 91.

For the sake of completeness we also tested other popular Monte Carlo Search algorithms. The results are given in Table 4 for UCT [25], Nested Monte Carlo Search [5,29] and Diversity NRPA [18]. The UCT constant is set to 0.4, NMCS is tested for repeated calls to level 1 and level 2. Diversity GNRPA [18] is called with a set of 5 sequences at level 1 and 1 sequence at level 2. The improved GNRPA algorithm gives better results than these algorithms.

Table 5 gives the results with time of the best parallel algorithm using different options. The Correction option is to fix a discrepancy between the NEMO paper and the NEMO code in the heuristic. The Order option is to order moves such as NEMO or to use the order of the string. Given the results in the table the two options do not seem to matter much.

Table 6 gives the number of problems solved with time using the root parallel GNRPA algorithm with 20 process. The results are slightly worse than when using leaf parallelization with 8 threads. It is due to problems that converge slowly and do not benefit from restarts and where leaf parallelization enables to improve during much longer than root parallelization the best policy. Given enough resources the best algorithm might be the combination of root and leaf parallelization as in [28]. The parallelization of NRPA proposed by Nagorko [27] is also appealing.

5   Conclusion

We experimented with various Monte Carlo Search algorithms for the Inverse Folding problem. We have used very limited domain knowledge, essentially using a small part of the NEMO heuristics for the bias. By applying general Monte Carlo Search heuristics we were able to solve as many problems as NEMO in comparable times.

Acknowledgment. Thanks to Fernando Portela for his NEMO program. This work was supported in part by the French government under management of Agence Nationale de la Recherche as part of the "Investissements d'avenir" program, reference ANR19-P3IA-0001 (PRAIRIE 3IA Institute).

References

1. Bonnet, É., Rząźewski, P., Sikora, F.: Designing RNA secondary structures is hard. J. Comput. Biol. **27**(3), 302–316 (2020)
2. Bouzy, B.: Monte-Carlo fork search for cooperative path-finding. In: Cazenave, T., Winands, M.H.M., Iida, H. (eds.) CGW 2013. CCIS, vol. 408, pp. 1–15. Springer, Cham (2014). https://doi.org/10.1007/978-3-319-05428-5_1
3. Bouzy, B.: Burnt pancake problem: new lower bounds on the diameter and new experimental optimality ratios. In: Proceedings of the Ninth Annual Symposium on Combinatorial Search, SOCS 2016, Tarrytown, NY, USA, 6–8 July 2016, pp. 119–120 (2016)
4. Browne, C., et al.: A survey of Monte Carlo tree search methods. IEEE Trans. Comput. Intell. AI Games **4**(1), 1–43 (2012). https://doi.org/10.1109/TCIAIG.2012.2186810
5. Cazenave, T.: Nested Monte-Carlo search. In: Boutilier, C. (ed.) IJCAI, pp. 456–461 (2009)
6. Cazenave, T.: Monte Carlo beam search. IEEE Trans. Comput. Intell. AI Games **4**(1), 68–72 (2012)
7. Cazenave, T.: Monte-Carlo expression discovery. Int. J. Artif. Intell. Tools **22**(1), 1250035 (2013)
8. Cazenave, T.: Nested rollout policy adaptation with selective policies. In: Cazenave, T., Winands, M.H.M., Edelkamp, S., Schiffel, S., Thielscher, M., Togelius, J. (eds.) CGW/GIGA -2016. CCIS, vol. 705, pp. 44–56. Springer, Cham (2017). https://doi.org/10.1007/978-3-319-57969-6_4
9. Cazenave, T.: Playout policy adaptation with move features. Theoret. Comput. Sci. **644**, 43–52 (2016)
10. Cazenave, T.: Generalized nested rollout policy adaptation. In: Monte Search at IJCAI (2020)
11. Cazenave, T., Jouandeau, N.: On the parallelization of UCT. In: Computer Games Workshop, Amsterdam, Netherlands (2007). https://hal.archives-ouvertes.fr/hal-02310186
12. Cazenave, T., Lucas, J.Y., Kim, H., Triboulet, T.: Monte Carlo vehicle routing. In: ATT at ECAI (2020)
13. Cazenave, T., Sevestre, J.B., Toulemont, M.: Stabilized nested rollout policy adaptation. In: Monte Search at IJCAI (2020)
14. Cazenave, T., Teytaud, F.: Application of the nested rollout policy adaptation algorithm to the traveling salesman problem with time windows. In: Hamadi, Y., Schoenauer, M. (eds.) LION 2012. LNCS, pp. 42–54. Springer, Heidelberg (2012). https://doi.org/10.1007/978-3-642-34413-8_4
15. Cazenave, T., Teytaud, F.: Beam nested rollout policy adaptation. In: Computer Games Workshop, ECAI 2012, pp. 1–12 (2012)
16. Chaslot, G.M.J.-B., Winands, M.H.M., van den Herik, H.J.: Parallel Monte-Carlo tree search. In: van den Herik, H.J., Xu, X., Ma, Z., Winands, M.H.M. (eds.) CG 2008. LNCS, vol. 5131, pp. 60–71. Springer, Heidelberg (2008). https://doi.org/10.1007/978-3-540-87608-3_6
17. Coulom, R.: Efficient selectivity and backup operators in Monte-Carlo tree search. In: van den Herik, H.J., Ciancarini, P., Donkers, H.H.L.M.J. (eds.) CG 2006. LNCS, vol. 4630, pp. 72–83. Springer, Heidelberg (2007). https://doi.org/10.1007/978-3-540-75538-8_7

18. Edelkamp, S., Cazenave, T.: Improved diversity in nested rollout policy adaptation. In: Friedrich, G., Helmert, M., Wotawa, F. (eds.) KI 2016. LNCS (LNAI), vol. 9904, pp. 43–55. Springer, Cham (2016). https://doi.org/10.1007/978-3-319-46073-4_4

19. Edelkamp, S., Gath, M., Cazenave, T., Teytaud, F.: Algorithm and knowledge engineering for the TSPTW problem. In: 2013 IEEE Symposium on omputational Intelligence in Scheduling (SCIS), pp. 44–51. IEEE (2013)

20. Edelkamp, S., Gath, M., Greulich, C., Humann, M., Herzog, O., Lawo, M.: Monte-Carlo tree search for logistics. In: Clausen, U., Friedrich, H., Thaller, C., Geiger, C. (eds.) Commercial Transport. LNL, pp. 427–440. Springer, Cham (2016). https://doi.org/10.1007/978-3-319-21266-1_28

21. Edelkamp, S., Gath, M., Rohde, M.: Monte-Carlo tree search for 3D packing with object orientation. In: Lutz, C., Thielscher, M. (eds.) KI 2014. LNCS (LNAI), vol. 8736, pp. 285–296. Springer, Cham (2014). https://doi.org/10.1007/978-3-319-11206-0_28

22. Edelkamp, S., Greulich, C.: Solving physical traveling salesman problems with policy adaptation. In: 2014 IEEE Conference on Computational Intelligence and Games (CIG), pp. 1–8. IEEE (2014)

23. Edelkamp, S., Tang, Z.: Monte-Carlo tree search for the multiple sequence alignment problem. In: Eighth Annual Symposium on Combinatorial Search (2015)

24. Kinny, D.: A new approach to the snake-in-the-box problem. In: ECAI, vol. 242, pp. 462–467 (2012)

25. Kocsis, L., Szepesvári, C.: Bandit based Monte-Carlo planning. In: Fürnkranz, J., Scheffer, T., Spiliopoulou, M. (eds.) ECML 2006. LNCS (LNAI), vol. 4212, pp. 282–293. Springer, Heidelberg (2006). https://doi.org/10.1007/11871842_29

26. Méhat, J., Cazenave, T.: Combining UCT and nested Monte Carlo search for single-player general game playing. IEEE Trans. Comput. Intell. AI Games 2(4), 271–277 (2010)

27. Nagorko, A.: Parallel nested rollout policy adaptation. In: IEEE Conference on Games (2019)

28. Negrevergne, B., Cazenave, T.: Distributed nested rollout policy for SameGame. In: Cazenave, T., Winands, M.H.M., Saffidine, A. (eds.) CGW 2017. CCIS, vol. 818, pp. 108–120. Springer, Cham (2018). https://doi.org/10.1007/978-3-319-75931-9_8

29. Portela, F.: An unexpectedly effective Monte Carlo technique for the RNA inverse folding problem. BioRxiv, p. 345587 (2018)

30. Poulding, S.M., Feldt, R.: Generating structured test data with specific properties using nested Monte-Carlo search. In: Genetic and Evolutionary Computation Conference, GECCO 2014, Vancouver, BC, Canada, 12–16 July 2014, pp. 1279–1286 (2014)

31. Poulding, S.M., Feldt, R.: Heuristic model checking using a Monte-Carlo tree search algorithm. In: Proceedings of the Genetic and Evolutionary Computation Conference, GECCO 2015, Madrid, Spain, 11–15 July 2015, pp. 1359–1366 (2015)

32. Rimmel, A., Teytaud, F., Cazenave, T.: Optimization of the nested Monte-Carlo algorithm on the traveling salesman problem with time windows. In: Di Chio, C., et al. (eds.) EvoApplications 2011. LNCS, vol. 6625, pp. 501–510. Springer, Heidelberg (2011). https://doi.org/10.1007/978-3-642-20520-0_51

33. Rosin, C.D.: Nested rollout policy adaptation for Monte Carlo Tree Search. In: IJCAI, pp. 649–654 (2011)

Monte Carlo Graph Coloring

Tristan Cazenave[✉], Benjamin Negrevergne, and Florian Sikora

Université Paris-Dauphine, PSL University, CNRS, LAMSADE, 75016 Paris, France
{Tristan.Cazenave,Benjamin.Negrevergne,Florian.Sikora}@dauphine.fr

Abstract. Graph Coloring is probably one of the most studied and famous problem in graph algorithms. Exact methods fail to solve instances with more than few hundred vertices, therefore, a large number of heuristics have been proposed. Nested Monte Carlo Search (NMCS) and Nested Rollout Policy Adaptation (NRPA) are Monte Carlo search algorithms for single player games. Surprisingly, few work has been dedicated to evaluating Monte Carlo search algorithms to combinatorial graph problems. In this paper we expose how to efficiently apply Monte Carlo search to Graph Coloring and compare this approach to existing ones.

1 Introduction

Given a graph G, a proper coloration of G consists in assigning a color to each vertex of the graph such that no adjacent vertices receive the same color. The chromatic number $\chi(G)$ of G is the minimum number of colors required to have a proper coloration for G. Determining the chromatic number of a graph is probably one of the most studied topics in graph algorithms and discrete mathematics. It has many applications, including scheduling, timetabling, or communication networks (see references in [24]). Unfortunately, identifying the chromatic number is notoriously hard to solve: it is already NP-hard even if the question is to decide if the graph can be colored with 3 colors, and it is essentially completely not-approximable [30].

To cope with this difficulty, the research community has tried a variety of different approaches: mathematical programming [24], exact moderately exponential algorithms [1], approximation algorithms on special graph classes [8], algorithms of parameterized complexity for structural parameters and data reduction [4,22,23,25], heuristics, meta-heuristics, etc. In practice exact methods generally fail to color graphs with more than few hundred vertices [24], so a large number of publications on graph coloring algorithms focus on the design and improvement of heuristics approaches.

Early heuristics for graph coloring were often based on pure local search strategies such as *TabuSearch* [19]. Nowadays, most efficient modern algorithms are still based a local search strategy, but they combine it with sophisticated exploration techniques to escape local minima (e.g. *Variable Neighborhood Search* [26] and *Variable Space Search* [20]). Building on a the idea of combining local search with more exploratory search procedures, Fleurent and Ferland

© Springer Nature Switzerland AG 2021
T. Cazenave et al. (Eds.): MCS 2020, CCIS 1379, pp. 100–115, 2021.
https://doi.org/10.1007/978-3-030-89453-5_8

have proposed to use the framework of hybrid algorithms which combine a local search operator with a population based algorithm [16]. This idea has inspired a lot of research in the field (see for example [17]), and ultimately led to the state-of-the-art algorithm HEAD [27] (*Hybrid Evolutionary Algorithm in Duet*).

In comparison with hybrid algorithms based on local search, very little work has been dedicated to evaluating the performance of Monte-Carlo for graph coloring problem (except [9]). This is probably because the idea of discovering highly constrained solutions through random sampling seems counter-intuitive at first. However, modern Monte-Carlo based algorithm naturally combine random search (which provides exploration) with a tree search driven by a stochastic policy learned during the search, (which help improving good local solutions). These two features make modern Monte-Carlo based algorithms good candidates for the graph coloring problem.

In this paper, we evaluate the performance of two Monte-Carlo based algorithm, *Nested Monte Carlo Search* (NMCS) [5] and *Nested Rollout Policy Adaptation* (NRPA) [29], for the graph coloring problem. As we will show, our modeling of the coloring problem as a Monte-Carlo search algorithm provides good performance and can compete with state-of-the-art hybrid algorithms which have been studied and improved over the past 30 years.

In Sect. 2, we review related work concerning Monte Carlo Search methods and describe in Sect. 3 the two we will use in this paper. In Sect. 4, we discuss various modeling choices for our approach. In Sect. 5 we describe the other algorithms for graph coloring that we will compare to ours. Finally, in Sect. 6, we conduct thorough experiments to demonstrate the performance of NRPA.

2   Monte Carlo Search Methods and Combinatorial Problems

Monte Carlo Tree Search algorithms (MCTS) have been most successful in the area of game artificial intelligence [3], and have obtained state-of-the-art in this field. They have also been applied to a variety of other problems in combinatorial optimization problems, but they remain marginally used in this area. For example, NRPA has been applied to the Traveling Salesman with Time Windows problem [7,10], and other applications also deal with 3D Packing with Object Orientation [12], the physical traveling salesman problem [13], the Multiple Sequence Alignment problem [14] or Logistics [11].

In 2017, Edelkamp and co-workers have applied Monte Carlo Search to graph coloring [9]. They compare various Monte-Carlo search algorithms such as NMCS, NRPA as well as a SAT-based approach. In their experiments they report that the best results were obtained with NMCS which contrast with our results in this paper. We propose to optimize further the modelling of the problem using node ordering, refined scoring and a different Adapt function. Our optimizations improve much the search time compare to the alternative modellings.

3 Nested Monte Carlo Search

In this section we describe the two Monte Carlo search algorithms that we have considered in the rest of this paper: *Nested Monte Carlo Search* (NMCS) and *Nested Rollout Policy Adaptation* (NRPA).

As most Monte-Carlo based algorithms, NMCS and NRPA produce a good solution by generating a large number of random sequences of branching decisions (a.k.a moves). The best sequence according to some objective function is then returned as a final solution to the problem. Since the quality of final sequence directly depends on the quality of the random sequences generated during the search, NMCS and NRPA combine a variety of techniques to improve the quality of the random sequence generator such as tree search, policy adaptation or nested algorithms.

At the lowest recursive level, the generation of random sequences is driven by a stochastic policy (a probability distribution over the moves). Random sequences are generated based on this policy by sampling moves from the policy using Gibbs sampling, as described in Algorithm 1. If we have access to background knowledge, it can be encoded as a non-uniform distribution over the moves in the policy. Otherwise, the initial stochastic policy assigns equal probability to each move.

Algorithm 1. The playout algorithm

---

playout (*state*, *policy*)
sequence ← []
while true **do**
  if *state* is terminal **then**
    return (score (*state*), *sequence*)
  end if
  $z ← 0.0$
  for m in possible moves for *state* **do**
    $z ← z + \exp (policy\ [m])$
  end for
  choose a move m with probability $\frac{exp(policy[m])}{z}$
  state ← play (*state*, m)
  sequence ← *sequence* + m
end while

In NMCS, the policy remains the same throughout the execution of the algorithm. However, the policy is combined with a tree search to improve the quality over a simple random sequence generator. At each step, each possible move is evaluated by completing the partial solution into a complete one using moves sampled from the policy. Whichever intermediate move has led to the best completed sequence, is selected and added to the current sequence. (See Algorithm 2.) The same procedure is repeated to choose the following move, until the sequence has reached a terminal state.

A major difference between NMCS and NRPA, is the fact that NRPA uses a stochastic policy that is *learned* during the search. At the beginning of the algorithm, the policy is initialized uniformly and later improved using gradient descent steps based the best sequence discovered so far (See. Algorithm 3). The procedure used to update the policy from a given sequence is given in Algorithm 4. Note that this difference is crucial because unlike NMCS, NRPA is able to *acquire* background knowledge about the problem being solved, and does not require the user to specify it. Ultimately, this knowledge will contribute to speed up the discovery of a good solution.

Algorithm 2. The NMCS algorithm.

NMCS (*state, level*)
if level == 0 **then**
   return playout (*state, uniform*)
end if
BestSequenceOfLevel ← ∅
while *state* is not terminal **do**
   for *m* in possible moves for *state* **do**
      s ← play (*state*, m)
      NMCS (*s, level* − 1)
      update *BestSequenceOfLevel*
   end for
   bestMove ← move of the *BestSequenceOfLevel*
   state ← play (*state, bestMove*)
end while

Finally, both algorithms are nested, meaning that at the lowest recursive level, weak random policies are used to sample a large number of low quality sequences, and produce a search policy of intermediate quality. At the recursive level above, this policy is used to produce sequence of high quality. This procedure is applied recursively, in general 4 or 5 times. In both algorithm the recursive level (denoted l) is a crucial parameter. Increasing l increases the quality of the final solution at the cost of more CPU time. In practice it is generally set to 4 or 5 recursive levels depending on the time budget and the computational resources available.

Algorithm 3. The NRPA algorithm.

```
NRPA (level, policy)
if level == 0 then
    return playout (root, policy)
end if
bestScore ← −∞
for N iterations do
    (result,new) ← NRPA(level − 1, policy)
    if result ≥ bestScore then
        bestScore ← result
        seq ← new
    end if
    policy ← Adapt (policy, seq)
end for
return  (bestScore, seq)
```

Algorithm 4. The Adapt algorithm

```
Adapt (policy, sequence)
polp ← policy
state ← root
for move in sequence do
    polp [move] ← polp [move] + α
    z ← 0.0
    for m in possible moves for state do
        z ← z + exp (policy [m])
    end for
    for m in possible moves for state do
        polp [m] ← polp [m] - α * exp(policy[m])/z
    end for
    state ← play (state, move)
end for
policy ← polp
```

4   Graph Coloring as a Monte Carlo Search Problem

In this section we discuss several alternative models to capture the Graph Coloring problem as a Monte Carlo Search problem. Remark that we focus *decision* problem (i.e. deciding if a graph can be colored with a given number of colors).

We start by defining the possible moves, and then present the node ordering heuristic, and deal with the question of generating valid moves. Finally, we discuss the objective function and as well as an optimization of the adapt function for the graph coloring problem.

4.1   Legal Moves

In the context of the graph coloring problem, a *move* consists in assigning a particular color to an uncolored vertex of the graph. Thus, given a graph $G = (V, E)$ and a set of colors C, a move is a pair (v, c) where v is an uncolored vertex in V, and c is any color in c.

NMCS and NRPA only consider *legal moves* at each step, and lowering the number of legal moves is a key performance issue for both algorithms. In NMCS, a large number of legal moves leads to a very large branching factor which slows down the algorithm at each recursive level. For NRPA a large number of legal moves results in a large policy vector, which makes it more difficult to train with comparatively less training examples.

A first naive approach consists in considering every possible move at every step, leading to a number of possible moves that can be as big as $|V| \times |C|$ in the initial condition. This solution results in poor efficiency and low quality solution which we do not report here. To lower the number of possible moves down, we adopt a different model in which each move vertex is considered in a particular order (e.g. random order). At each step, only one node and all its legal coloration are considered. This reduces the maximum number of moves from $|V| \times |C|$ to $|C|$. As we will see, it leads to good results in practice. However, imposing an order over the vertices induces a strong bias over the exploration of the search space, which we study in the next section.

4.2   Node Order

The naive approach to node ordering is to fix a predefined or random order of the nodes and to color them in this order. A better heuristic is DSatur [2]. It chooses as the next node to color the node that has the less possible colors. In case multiple nodes have the same minimal number of possible colors it break ties by choosing the node that has the most neighbors. DSatur is a good heuristic to order nodes for NRPA since it propagates the constraints in the graph and avoids choosing colors for a node that would reveal inconsistent later due to more constrained neighbors. For example if a node has only one possible color it will always be chosen first by DSatur. By doing so the neighboring nodes have one less possible color and it avoids taking this impossible colors for neighboring nodes which would not have been the case if the one color node had been chosen later.

4.3   Selective Search

When trying possible colors for a node it is not wise to choose a color that is already assigned to a neighboring node. In order to avoid as much as possible bad branching decisions we use forward checking. When selecting the color for a node, all the colors of the neighboring nodes are removed from the set of possible colors for the node. This is related to selective NRPA [6] where heuristics are used to avoid bad moves. However in our case of Graph Coloring the moves that are discarded are moves that can never be part of a valid solution. So it is safe to remove them. Inconsistent colors are never considered as possible moves except if there a no possible color for a node since neighboring nodes already contains all the available colors. In this case all colors are considered possible and the algorithm chooses a color for the node even if it is inconsistent.

4.4 Scoring Function

Another design choice is the way to score a playout. In [5] the depth of the playout was used for Sudoku and the playouts were stopped when reaching an inconsistent state, i.e. a state where a variable has no more possible values. For Graph Coloring we use a more informed score. We count the number of inconsistent edges, i.e. monochromatic edges. If there are two adjacent vertices with the same color, the score decrease by one. A score equal to the number of edges of the graph means that we have found a solution. Note that we also tried a scoring function mixing both the number of colors and the number of inconsistent edges (trying to decrease both), which would allow the algorithm to solve the optimization problem directly, but it didn't work well.

We also experimented with NMCS. For the sake of completeness NMCS is given in Algorithm 2. The score of the playouts, the selection of edges, and the selectivity of colors in NMCS are the same as in NRPA.

4.5 Coding the Moves

In NRPA it is important to design how moves are coded. There is a bijection between moves and integer such that moves are associated to weights. We choose to use a simple coding for our moves: the index of a node multiplied by the number of colors plus the index of the color in the move.

4.6 Adapt All the Colors

When modifying the weights with the adapt function, there are two options. The first one is to modify the weights of the possible moves and to adapt using only the probabilities of the moves that can be played in the current state. The second one is to modify the weights for all the colors, including the colors that were discarded as possible moves since they were inconsistent with neighboring nodes.

The standard Adapt function is given in Algorithm 4. The modified function that modifies the probabilities for all the colors is given in Algorithm 5.

Algorithm 5. The AdaptAll algorithm

AdaptAll (*policy, sequence*)
$polp \leftarrow policy$
$state \leftarrow root$
for *move* in *sequence* **do**
  $polp$ [code(*move*)] $\leftarrow polp$ [code(*move*)] $+ \alpha$
  $z \leftarrow 0.0$
  for m in all possible colors even the illegal ones **do**
    $z \leftarrow z + \exp (policy$ [code(m)])
  end for
  for m in all possible colors even the illegal ones **do**
    $polp$ [code(m)] $\leftarrow polp$ [code(m)] $- \alpha * \frac{exp(policy[code(m)])}{z}$
  end for
  $state \leftarrow$ play (*state, move*)
end for
$policy \leftarrow polp$

5   Compared Approaches for Graph Coloring

In this section we present the other Graph Coloring algorithms that we used to compare with our approach.

5.1   SAT

We used the following SAT encoding to decide if one can color a graph $G = (V, E)$ with k colors (this formulation is for example used in [21]). We add a variable $x_{v,i}$ for each $v \in V$ and each $i \in [k]$. Then, for each $v \in V$, we add a clause $(\bigvee_i x_{v,i})$, ensuring that each vertex receives a color, and for each edge $uv \in E$ and each color $i \in [k]$, we add a clause $(\neg x_{u,i} \vee \neg x_{v,i})$ such that adjacent vertices receives different colors. This formula is true iff there is a k-coloration of G. Note that if there is no truth assignment for the formula, it tells that the graph is not k-colorable. However, we will not use this in our experiments.

As a solver, we used MiniSat to solve the built formula [15].

5.2   HEAD

HEAD [27] is an hybrid meta-heuristic, more precisely a memetic algorithm, mixing a local search procedure (Tabu-Search) with an evolutionary algorithm. It is based on the *Hybrid Evolutionary Algorithm* (HEA) by Galinier and Hao [17].

The general principle of HEA and HEAD is to start with a population of individuals, which are first improved using a local search procedure. Then, a crossover operator is applied to the best individuals in order to generate new diverse individuals and the procedure is repeated for a fixed number of steps which are called generations.

However general crossover operators do not work well for the graph coloring problem, so the main contribution of HEA is a specialized crossover operator. In HEA each individual is a partition of vertices into color classes, and the crossover operator is required to preserve color classes or subset of color classes.

HEAD builds on HEA by introducing various improvements including an original method to maintain diversity inside the population: individuals from earlier generations are re-introduced as candidate individuals in later generations. Using these technique, HEAD has been able to rediscover known coloring at much lower computational cost that earlier approaches [27].

The source-code of HEAD is available online[1]. Authors provided experiments showing good performances on the classical benchmarks and is probably the faster heuristic to date.

5.3   Greedy Coloring

For greedy coloring we use the same generation of possible moves as NRPA except that we only play one playout and that the maximum numbers of colors is not fixed. The order in which vertices are visited during this greedy coloring is the one given by DSatur [2]. Greedy Coloring is used to establish initial upper bounds for NRPA, NMCS, SAT and HEAD that are lowered down using search to establish better upper bounds.

6   Experiments

In this section, we compare the performance of two Monte-Carlo based approaches described in Sect. 3 and 4 with the other approaches described in Sect. 5.

6.1   Experimental Protocol

Execution Strategy: In practice we observe a high variance in runtimes and best results throughout the different runs of the same algorithm. To reduce the variance in the results, and allow a fair comparison, we proceed as follows: each algorithm is executed 5 times for a given number of color k with a timeout of 30 min. If the algorithm discovers at least one valid k-coloring for the graph instance, we decrease k by one, and repeat this procedure until the algorithm is unable to discover a k-coloring within the timeout limit. Then, we report the lowest k for which the algorithm was able to discover a coloring (denoted **UB** in the result tables), as well as the success rate for this lowest k (denoted **Reached**). The initial value of k to start with is determined with the simple greedy algorithm with nodes ordered according to DSatur (denoted **UBI**). Sometimes the algorithm is unable to improve over the simple greedy algorithm, which we signal with a '–' in the result table (unless the greedy algorithm has already discovered the minimum number χ).

---

[1] https://github.com/graphcoloring/HEAD/.

Test Instances: We used standard benchmark instances available on the website maintained by Gualandi and Chiarandini [18], collected from DIMACS benchmark. This benchmark has been used extensively to evaluate graph coloring algorithms and is now considered to be the standard set for experimentation [24] in this field. Moreover, because these instances have been extensively studied, the optimal chromatic number χ is known for most of them.

These instances are sorted by difficulty. Instances marked NP-m (resp. NP-d) should be solved in less than an hour (resp. than a day). For the harder instances marked NP-?, either the chromatic number is unknown or the time needed to solve them is unknown to [18].

Hardware and Implementations Details: Every execution reported in this experimental section has been conducted on a Intel Xeon E5-2630 v3 (Haswell, 2.40 GHz). Although some algorithm support parallel execution, (e.g. NRPA and HEAD), we on report execution times on sequential execution (using one thread) to reduce the variance in execution times, and to allow meaningful comparison with other purely sequential algorithms.

The peak memory usage is limited to 4 GB which is not a limitation for any of the algorithm except for the SAT model which runs out of memory sometimes.

We implemented algorithms using SAT, NRPA and NMCS in C++, using the Boost Graph Library. For NRPA, we use our implementation described in [28] and available online[2]. In our experiments we use $l = 7, N = 100$ for NRPA[3]. For NMCS, we use increasing nesting levels. This mean that we start the search with a level of 1, and if no solution is found, we increment the level and repeat.

Table 1. Results for the easy instances (marked NP-m) with a timeout of 30 min.

Instance	$|V|$	$|E|$	χ	UBI	NMCS		NRPA		SAT		HEAD	
					UB	Reached	UB	Reached	UB	Reached	UB	Reached
1-FullIns_4	93	593	5	5	5	100%	5	100%	5	100%	5	100%
2-FullIns_4	212	1621	6	6	6	100%	6	100%	6	100%	6	100%
3-FullIns_3	80	346	6	6	6	100%	6	100%	6	100%	6	100%
4-FullIns_3	114	541	7	7	7	100%	7	100%	7	100%	7	100%
5-FullIns_3	154	792	8	8	8	100%	8	100%	8	100%	8	100%
ash608GPIA	1216	7844	4	6	4	100%	4	100%	4	100%	4	100%
ash958GPIA	1916	12506	4	6	4	60%	4	100%	4	100%	4	100%
le450_15a	450	8168	15	17	15	60%	15	100%	15	100%	15	100%
mug100_1	100	166	4	4	4	100%	4	100%	4	100%	4	100%
mug100_25	100	166	4	4	4	100%	4	100%	4	100%	4	100%
qg.order40	1600	62400	40	42	40	100%	40	100%	40	100%	40	100%
wap05a	905	43081	50	50	50	100%	50	100%	50	100%	50	100%
myciel6	95	755	7	7	7	100%	7	100%	7	100%	7	100%
school1_nsh	352	14612	14	26	14	100%	14	100%	14	100%	14	100%
Avg. ratio to χ					1.0000		1.0000		1.0000		1.0000	

---

[2] https://github.com/bnegreve/nrpa.

[3] Note that level 7 will probably never be reached in a reasonable amount of time, this is to allow NRPA to continue to search until a solution is found; this value of N gave often better results than smaller or bigger values.

Table 2. Results for the instances marked NP-h by [18] with a timeout of 30 min.

Instance	$|V|$	$|E|$	χ	UBI	NMCS		NRPA		SAT		HEAD	
					UB	Reached	UB	Reached	UB	Reached	UB	Reached
flat300_28_0	300	21695	28	41	38	20%	35	20%	39	100%	**31**	100%
r1000.5	1000	238267	234	248	243	20%	**240**	40%	247	100%	248	–
r250.5	250	14849	65	67	**65**	100%	**65**	100%	**65**	100%	66	40%
DSJR500.5	500	58862	122	132	125	60%	**122**	40%	126	100%	124	60%
DSJR500.1c	500	121275	85	88	88	–	87	60%	**86**	100%	**86**	80%
DSJC125.5	125	3891	17	23	19	100%	18	100%	19	100%	**17**	100%
DSJC125.9	125	6961	44	50	45	40%	**44**	100%	46	100%	**44**	100%
DSJC250.9	250	27897	72	90	84	20%	76	20%	86	100%	**72**	100%
queen10_10	100	2940	11	14	**11**	60%	**11**	40%	12	100%	**11**	100%
queen11_11	121	3960	11	14	13	100%	13	100%	13	100%	**12**	100%
queen12_12	144	5192	12	16	14	100%	14	100%	14	100%	**13**	100%
queen13_13	169	6656	13	17	15	100%	15	100%	15	100%	**14**	100%
queen14_14	196	4186	14	19	16	100%	16	100%	16	100%	**15**	100%
queen15_15	225	5180	15	20	17	40%	17	100%	18	100%	**16**	100%
Avg. ratio to χ					1.1101		1.0851		1.1276		1.0428	

6.2   Results

We give results of Monte Carlo approaches (NMCS and NRPA) compared to other approaches in Tables 1,2,3 and 4.

Table 3. Results for the difficult instances marked NP-? by [18] with a timeout of 30 min.

Instance	$|V|$	$|E|$	χ	UBI	NMCS		NRPA		SAT		HEAD	
					UB	Reached	UB	Reached	UB	Reached	UB	Reached
le450_5a	450	5714	5	10	6	20%	**5**	100%	**5**	100%	**5**	100%
le450_5b	450	5734	5	7	6	40%	**5**	20%	**5**	100%	**5**	100%
le450_15b	450	8169	15	17	**15**	100%	**15**	100%	**15**	100%	**15**	100%
le450_15c	450	16680	15	24	22	100%	21	100%	22	100%	**15**	100%
le450_15d	450	16750	15	24	22	100%	20	20%	22	100%	**15**	100%
le450_25c	450	17343	25	28	27	100%	**26**	100%	27	100%	**26**	100%
le450_25d	450	17425	25	29	27	100%	**26**	100%	27	100%	**26**	100%
qg.order60	3600	212400	60	63	**60**	40%	62	100%	61	100%	**60**	100%
qg.order100	10000	990000	100	106	–	20%	102	20%	–	20%	**100**	100%
Avg. ratio to χ					1.7626		1.0963		1.1300		1.0089	

Table 4. Results for the very difficult problems (NP-?) with a timeout of 30 min. The chromatic number of these graphs seems unknown and we only know a lower bound via [18].

Instance	$|V|$	$|E|$	χ_{LB}	UBI	NMCS UB	NMCS Reached	NRPA UB	NRPA Reached	SAT UB	SAT Reached	HEAD UB	HEAD Reached
DSJC250.1	250	3218	4	10	9	100%	**8**	40%	9	100%	**8**	100%
DSJC250.5	250	15668	26	37	34	100%	32	100%	35	100%	**28**	100%
DSJC500.1	500	12458	9	16	14	40%	14	100%	15	100%	**12**	100%
DSJC500.5	500	62624	43	65	62	20%	59	80%	63	100%	**48**	100%
DSJC500.9	500	112437	123	163	161	20%	148	20%	163	–	**126**	100%
DSJC1000.1	1000	49629	10	25	25	–	24	100%	25	–	**21**	100%
DSJC1000.5	1000	249826	73	114	114	–	112	40%	114	–	**83**	60%
DSJC1000.9	1000	449449	216	301	301	–	299	40%	301	–	**223**	20%
flat1000_50_0	1000	245000	15	113	113	–	111	40%	113	–	**50**	100%
flat1000_60_0	1000	245830	14	112	112	–	112	–	112	–	**60**	100%
flat1000_76_0	1000	246708	14	115	115	–	110	20%	113	100%	**82**	80%
r1000.1c	1000	485090	96	107	107	–	107	–	105	100%	**98**	20%
abb313GPIA	1557	53356	8	11	11	–	11	–	**9**	100%	**9**	60%
latin_square_10	900	307350	90	129	129	–	121	20%	129	–	**103**	20%
wap01a	2368	110871	41	47	46	60%	45	40%	43	100%	**42**	60%
wap02a	2464	111742	40	46	46	–	45	100%	**42**	100%	**42**	100%
wap03a	4730	286722	40	57	55	40%	55	100%	46	100%	**44**	20%
wap04a	5231	294902	40	46	46	–	46	–	**44**	100%	**44**	100%
wap06a	947	43571	40	44	42	100%	**41**	100%	**41**	100%	42	60%
wap07a	1809	103368	40	47	44	80%	43	60%	**42**	100%	**42**	100%
wap08a	1870	104176	40	44	43	20%	43	100%	**42**	100%	**42**	100%
C2000.5	2000	999836	99	207	207	–	207	–	207	–	**151**	40%
C4000.5	4000	4000268	107	376	376	–	376	–	376	–	**281**	20%
Avg. ratio to χ					2.3746		2.3173		2.3410		1.7027	

First, we observe that the simple greedy algorithm is generally able to discover the χ value for a number of instances in Table 1. When it is not the case, all the approaches we discuss in this paper have been able to tackle these instances, and discover the χ. However, NMCS does not reach 100% success rate on two of these instances.

When we look at the other results, we can see that NMCS is generally the weakest algorithm, and obtains the worst (higher) ratio to χ for the very difficult instances in Table 3 and 4, often significantly worse than the ratio for NRPA. Note that with a different representation of the problem, authors of [9] reported better results with NMCS, contrary to our results. We also observe that the performance gap between the two approaches is small on the medium NP-m instances (Table 1), but large on the most difficult instances (Table 3). This suggests that with our modeling, NRPA is able to acquire better policies along the execution of the algorithm. The benefit of the learned policies over the tree search becomes more visible in long runs on difficult instances. This motivates the general idea of introducing learning into Monte-Carlo search in order to improve the quality of the search.

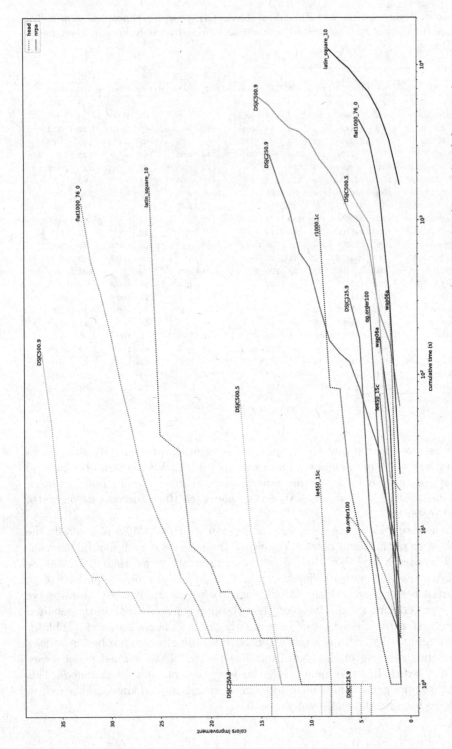

Fig. 1. Comparison of computation times (log scale) between NRPA (solid lines) and HEAD (dotted lines) for some instances.

We can also see that NRPA is generally better than the SAT model, but SAT is surprisingly good for the difficult wap instances in Table 4, unfortunately, we are unable to explain this result and further experiments are needed.

However, a pair-wise comparison between the upper bounds discovered by NRPA and the ones discovered by HEAD demonstrates that HEAD is better in general. It is worth mentioning that algorithms such as HEAD include specialized graph coloring operators which have been extensively studied over the past decades. In contrast our algorithm is based on a general purpose implementation of NRPA, and only includes the specializations we have described in Sect. 4. Nevertheless, NRPA is the second best in all the datasets, and is often as good, or even better than HEAD.

In Fig. 1 we further analyze the behaviour of the two algorithms by comparing the execution times. We selected the best of 5 runs for each tested number of colors and for each algorithm. The time is the accumulated time, starting with the number of colors computed by the greedy algorithm. The y-axis represents the number of improvement by an algorithm for this instance starting from the greedy coloring (i.e. a value of 3 means that the algorithm gave a coloration with 3 less colors than the greedy algorithm did). For readability, we didn't include instances like C4000.5, where HEAD is very good. It seems that HEAD improves quite fast the number of colors but struggles to improve more over time (this is not always true however), while NRPA seems to benefit of longer running time. Indeed, there is not much improvement after 2000s for HEAD while NRPA continues to find colorations after 3000 s. This could be because HEAD is stuck in local optimum and cannot find new solutions, while NRPA continues to improve policies while exploring the search space.

This demonstrate that Monte Carlo based algorithms have the ability to compete with state-of-the-art hybrid algorithms on the graph coloring problem, and deserve further investigation with more optimizations (more specific strategies, restarts...). From a broader perspective, this also shows that a continuous optimization algorithms can be used to solve discrete problems such as the graph coloring problem.

7   Conclusion

In this paper, we deepen our understanding of Monte Carlo Search algorithms applied to Graph Coloring. Our method is significantly different from most other methods from the literature, and yet, it is able to compete with state-of-the-art algorithms which have been intensively optimized during the past decades. These results suggest that Monte Carlo search combined with policy adaptation are able to explore the search to discover good, yet diverse solutions. And that this technique should be investigated further alongside with more standard approaches.

It would be also interesting to see if NMCS could combine with a good heuristic like HEAD, by either using the local search to improves the solution or by using the genetic algorithm as playouts.

Future works includes the use of Graph Convolution Networks to model more complex policies that can make branching decisions based on the structure of graph at hand, and generalize knowledge from one graph to another.

Finally, we could apply Monte Carlo methods to other variants of graph coloring, like for example Weighted Vertex Coloring or Minimum Sum Coloring, since only the evaluation function would change.

Acknowledgment. This work was supported in part by the French government under management of Agence Nationale de la Recherche as part of the "Investissements d'avenir" program, reference ANR19-P3IA-0001 (PRAIRIE 3IA Institute).

References

1. Björklund, A., Husfeldt, T., Koivisto, M.: Set partitioning via inclusion-exclusion. SIAM J. Comput. **39**(2), 546–563 (2009)
2. Brélaz, D.: New methods to color the vertices of a graph. Commun. ACM **22**(4), 251–256 (1979)
3. Browne, C., et al.: A survey of Monte Carlo tree search methods. IEEE Trans. Comput. Intell. AI Games **4**(1), 1–43 (2012)
4. Cai, L.: Parameterized complexity of vertex colouring. Discret. Appl. Math. **127**(3), 415–429 (2003)
5. Cazenave, T.: Nested Monte-Carlo search. In: Boutilier, C. (ed.) IJCAI, pp. 456–461 (2009)
6. Cazenave, T.: Nested rollout policy adaptation with selective policies. In: Cazenave, T., Winands, M.H.M., Edelkamp, S., Schiffel, S., Thielscher, M., Togelius, J. (eds.) CGW/GIGA -2016. CCIS, vol. 705, pp. 44–56. Springer, Cham (2017). https://doi.org/10.1007/978-3-319-57969-6_4
7. Cazenave, T., Teytaud, F.: Application of the nested rollout policy adaptation algorithm to the traveling salesman problem with time windows. In: Hamadi, Y., Schoenauer, M. (eds.) LION 2012. LNCS, pp. 42–54. Springer, Heidelberg (2012). https://doi.org/10.1007/978-3-642-34413-8_4
8. Demaine, E.D., Hajiaghayi, M.T., Kawarabayashi, K.: Algorithmic graph minor theory: decomposition, approximation, and coloring. In: 46th Annual IEEE Symposium on Foundations of Computer Science (FOCS 2005), Pittsburgh, PA, USA, 23–25 October 2005, Proceedings, pp. 637–646. IEEE Computer Society (2005)
9. Edelkamp, S., Externest, E., Kühl, S., Kuske, S.: Solving graph optimization problems in a framework for Monte-Carlo search. In: Tenth Annual Symposium on Combinatorial Search (2017)
10. Edelkamp, S., Gath, M., Cazenave, T., Teytaud, F.: Algorithm and knowledge engineering for the TSPTW problem. In: 2013 IEEE Symposium on Computational Intelligence in Scheduling (SCIS), pp. 44–51. IEEE (2013)
11. Edelkamp, S., Gath, M., Greulich, C., Humann, M., Herzog, O., Lawo, M.: Monte-Carlo tree search for logistics. In: Clausen, U., Friedrich, H., Thaller, C., Geiger, C. (eds.) Commercial Transport. LNL, pp. 427–440. Springer, Cham (2016). https://doi.org/10.1007/978-3-319-21266-1_28
12. Edelkamp, S., Gath, M., Rohde, M.: Monte-Carlo tree search for 3D packing with object orientation. In: Lutz, C., Thielscher, M. (eds.) KI 2014. LNCS (LNAI), vol. 8736, pp. 285–296. Springer, Cham (2014). https://doi.org/10.1007/978-3-319-11206-0_28

13. Edelkamp, S., Greulich, C. :Solving physical traveling salesman problems with policy adaptation. In: 2014 IEEE Conference on Computational Intelligence and Games (CIG), pp. 1–8. IEEE (2014)
14. Edelkamp, S., Tang, Z.: Monte-Carlo tree search for the multiple sequence alignment problem. In: Eighth Annual Symposium on Combinatorial Search (2015)
15. Eén, N., Sörensson, N.: An extensible SAT-solver. In: Giunchiglia, E., Tacchella, A. (eds.) SAT 2003. LNCS, vol. 2919, pp. 502–518. Springer, Heidelberg (2004). https://doi.org/10.1007/978-3-540-24605-3_37
16. Fleurent, C., Ferland, J.A.: Genetic and hybrid algorithms for graph coloring. Ann. Oper. Res. **63**(3), 437–461 (1996). https://doi.org/10.1007/BF02125407
17. Galinier, P., Hao, J.-K.: Hybrid evolutionary algorithms for graph coloring. J. Comb. Optim. **3**(4), 379–397 (1999). https://doi.org/10.1023/A:1009823419804
18. Gualandi, S., Chiarandini, M.: Graph coloring benchmarks. https://sites.google.com/site/graphcoloring/vertex-coloring. Accessed 17 Nov 2019
19. Hertz, A., de Werra, D.: Using tabu search techniques for graph coloring. Computing **39**(4), 345–351 (1987). https://doi.org/10.1007/BF02239976
20. Hertz, A., Plumettaz, M., Zufferey, N.: Variable space search for graph coloring. Discret. Appl. Math. **156**(13), 2551–2560 (2008)
21. Ignatiev, A., Morgado, A., Marques-Silva, J.: Cardinality encodings for graph optimization problems. In: Sierra, C. (ed.) Proceedings of the Twenty-Sixth International Joint Conference on Artificial Intelligence, IJCAI 2017, Melbourne, Australia, 19–25 August 2017, pp. 652–658. ijcai.org (2017)
22. Jansen, B.M.P., Kratsch, S.: Data reduction for graph coloring problems. Inf. Comput. **231**, 70–88 (2013)
23. Lampis, M.: Finer tight bounds for coloring on clique-width. In: Chatzigiannakis, I., Kaklamanis, C., Marx, D., Sannella, D. (eds.) 45th International Colloquium on Automata, Languages, and Programming, ICALP 2018, Prague, Czech Republic, 9–13 July 2018, vol. 107 of LIPIcs, pp. 86:1–86:14. Schloss Dagstuhl - Leibniz-Zentrum fuer Informatik (2018)
24. Malaguti, E., Toth, P.: A survey on vertex coloring problems. ITOR **17**(1), 1–34 (2010)
25. Marx, D.: Parameterized coloring problems on chordal graphs. Theor. Comput. Sci. **351**(3), 407–424 (2006)
26. Mladenović, N., Hansen, P.: Variable neighborhood search. Comput. Oper. Res. **24**(11), 1097–1100 (1997)
27. Moalic, L., Gondran, A.: Variations on memetic algorithms for graph coloring problems. J. Heuristics **24**(1), 1–24 (2017). https://doi.org/10.1007/s10732-017-9354-9
28. Negrevergne, B., Cazenave, T.: Distributed nested rollout policy for SameGame. In: Cazenave, T., Winands, M.H.M., Saffidine, A. (eds.) CGW 2017. CCIS, vol. 818, pp. 108–120. Springer, Cham (2018). https://doi.org/10.1007/978-3-319-75931-9_8
29. Rosin, C.D.: Nested rollout policy adaptation for Monte Carlo tree search. In: IJCAI, pp. 649–654 (2011)
30. Zuckerman, D.: Linear degree extractors and the inapproximability of max clique and chromatic number. Theory Comput. **3**(1), 103–128 (2007)

Enhancing Playout Policy Adaptation for General Game Playing

Chiara F. Sironi[1], Tristan Cazenave[2]([⊠]), and Mark H. M. Winands[1]

[1] Game AI & Search Group, Department of Data Science and Knowledge
Engineering, Maastricht University, Maastricht, The Netherlands
{c.sironi,m.winands}@maastrichtuniversity.nl
[2] LAMSADE, Université Paris-Dauphine, PSL, CNRS, Paris, France
tristan.cazenave@lamsade.dauphine.fr

Abstract. Playout Policy Adaptation (PPA) is a state-of-the-art strategy that has been proposed to control the playouts in Monte-Carlo Tree Search (MCTS). PPA has been successfully applied to many two-player, sequential-move games. This paper further evaluates this strategy in General Game Playing (GGP) by first reformulating it for simultaneous-move games. Next, it presents five enhancements for the strategy, four of which have been previously successfully applied to a related MCTS playout strategy, the Move-Average Sampling Technique (MAST). Experiments on a heterogeneous set of games show three enhancements to have a positive effect on PPA: (i) updating the policy for all players proportionally to their payoffs instead of updating only the policy of the winner, (ii) collecting statistics for N-grams of moves instead of single moves only, and (iii) discounting the backpropagated payoffs depending on the depth of the playout. Results also show enhanced PPA variants to be competitive with MAST for small search budgets, and better for larger search budgets. The use of an ϵ-greedy selection of moves and of after-move decay of statistics, instead, seem to have a detrimental effect on PPA.

Keywords: Monte-Carlo Tree Search · Playout policy adaptation · General game playing

1 Introduction

Monte-Carlo Tree Search (MCTS) [8,15] is a simulation-based search algorithm that has been successfully applied to various domains [2], one of which is General Game Playing (GGP) [13]. In GGP, the aim is to create agents that are able to play a wide variety of possibly unknown games, only by being given their rules at run-time. The fact that MCTS does not necessarily need domain-specific knowledge is what makes it suitable to be used in GGP.

Many successful enhancements and modifications of MCTS have been investigated [2]. In GGP, particular attention has been given to strategies that can

© Springer Nature Switzerland AG 2021
T. Cazenave et al. (Eds.): MCS 2020, CCIS 1379, pp. 116–139, 2021.
https://doi.org/10.1007/978-3-030-89453-5_9

learn how to control the search on-line, i.e. while playing the game, because no domain-specific knowledge is available in advance. Playout Policy Adaptation (PPA) [3] is among the domain-independent strategies that have been proposed to guide the search in MCTS. PPA learns the playout policy on-line by keeping weights for each move visited during the search and adapting them according to the performance of the moves during game simulations. The weights of the moves performed by the winner of a simulation are increased, while the weights of all the legal moves that were available for the winner are decreased. The learned weights are then used in each playout to define a move-selection policy. PPA has been successfully applied to various two-player, sequential-move games [3]. This, together with the fact that it is domain-independent, makes it promising to be further evaluated in GGP.

Moreover, PPA shares some characteristics with another domain-independent playout strategy for MCTS, the Move-Average Sampling Technique (MAST) [11]. Like PPA, MAST collects statistics about the general performance of the available moves in the game, and biases the playout towards moves that have performed overall well so far. The main difference is that MAST learns the value of a move by looking only at its average performance in the game, while PPA learns it by comparing the move with the other available moves for the state in which it has been played. For MAST, various successful modifications have been tested in the literature [11, 23, 24]. The similarity of MAST and PPA, makes such modifications promising to be tested for the latter as well.

The main goal of this paper is to evaluate the PPA algorithm for General Game Playing, looking at possible enhancements that can further improve its performance. First, this paper extends the formulation of the PPA algorithm for simultaneous-move games to make it applicable also to such type of games. Moreover, five different modifications of the PPA algorithm are presented. One of them consists in updating the statistics of all players proportionally to their payoff as MAST does, instead of updating only the statistics of the winner by a fixed amount. Three of them are enhancements previously tested successfully for MAST [11, 23, 24]. More precisely, the ϵ-greedy selection of moves, the decay of statistics in-between subsequent game steps, and the use of statistics collected for N-grams (i.e. sequences) of moves, instead of only for single moves. Finally, the fifth modification consist in considering information about the length of the simulation when updating the playout policy, in order to bias the playout towards moves that win quickly. This modification is implemented also for MAST, to evaluate how it influences a playout strategy similar to PPA. The PPA strategy and its enhancements are tested on a heterogeneous set of games of the Stanford GGP project [12], of which the rules are represented in GDL (Game Description Language) [16].

The paper is structured as follows. First, Sect. 2 presents previous work related to PPA. Next, Sect. 3 gives background knowledge on MCTS, MAST and its enhancements. Subsequently, Sects. 4 and 5 describe the PPA algorithm and its enhancements, respectively. Experimental results are discussed in Sect. 6. Finally, Sect. 7 gives the conclusions and discusses future work.

2   Related Work

The PPA algorithm seems to be a promising playout strategy for MCTS in many domains. Previous literature already investigated variants of the algorithm that improve its performance on a set of two-player, sequential-move games (e.g. Breakthough, Knightthrough, Domineering, ...). For example, Cazenave [4] proposed to extend PPA with move features (PPAF). The PPAF algorithm learns weights like PPA does, but distinguishes moves depending on some features they are associated with (e.g. whether they are a capture move or a simple movement). PPAF is harder to apply in GGP, because it would require an on-line analysis of the game rules to first extract relevant move features for a given game that might be previously unknown. Another example of PPA variant that has been investigated is the PPAF strategy with memorization (PPAFM) [5]. With respect to PPAF, the PPAFM strategy does not reset the learned policy between game steps, but keeps it memorized and reuses it. This modification can be easily applied to standard PPA as well.

The interest in the PPA algorithm is motivated also by the success of the closely related Nested Rollout Policy Adaptation (NRPA) [18]. NRPA is designed for single-player games and, similarly to PPA, it learns a playout policy by associating weights to the different available moves. The difference is that NRPA is structured with nested search levels and updates the policy using the best playout found so far at each level. Because the best playout is ill-defined for multi-player games, PPA updates the policy of a given player using the playouts that resulted in a win for such player. NRPA has achieved good results in games, such as Morpion Solitaire and crossword puzzles [18], but has also been applied to solve traveling salesman problems [7,9] and logistic problems [10], such as vehicle routing, container packing and robot motion planning.

Finally, as already mentioned, PPA is similar to the MAST strategy, which has been successfully applied to GGP [11] and has been subsequently extended to the N-grams Selection Technique (NST) [23]. NST uses statistics of N-grams of moves to guide the playouts and has been shown successful as domain-independent strategy in GGP both for board games [23] as well as for video games [22]. Moreover, Powley et al. [17] used the same idea of NST to implement their strategy, the N-gram-Average Sampling Technique (NAST). This technique has been shown successful also for imperfect-information games [17].

3   Background

This section provides background on MCTS (Subsect. 3.1), on the UCT selection strategy (Subsect. 3.2), on the MAST playout strategy (Subsect. 3.3), and on the NST playout strategy (Subsect. 3.4).

3.1   Monte-Carlo Tree Search

MCTS [8,15] is a simulation based-search strategy that incrementally builds a tree representation of the state space of a game. MCTS performs multiple simulations of the game until a given search budget expires. Each MCTS simulation is divided into the following four phases:

Selection: during this phase, MCTS visits the tree built so far using a *selection strategy* to choose which move to visit in each traversed node. The selection phase ends when a tree node that has not been fully expanded is reached. A node is fully expanded when all its successor nodes have been added to the tree.

Expansion: during this phase, one or more nodes are added to the tree. Such node(s) are selected among the not yet explored children of the last node visited during the selection phase.

Playout: during the playout phase, MCTS starts from a node that was added to the tree during the expansion phase, and simulates the game until a terminal state or a fixed depth is reached. A *playout strategy* is used in each state to select which move to visit. The simplest playout strategy selects moves randomly among the legal ones.

Backpropagation: during this phase, the payoff obtained at the end of the simulation is propagated back through all the visited tree nodes, and used to update statistics about the performed moves. For multi-player games, a tuple of payoffs, one for each player, is propagated back in the tree.

When the search budget expires, MCTS returns one of the moves in the root node to be played in the real game. Commonly, the move that is returned is either the one with the highest number of visits or the one with the highest average payoff. This paper uses the second option.

Note that for simultaneous-move games, MCTS visits in each state a joint move. This paper uses an implementation of MCTS that represents joint moves as vectors formed by one individual move for each player, as described by Sironi [21]. This means that both the selection strategy and the playout strategy have to select a move for each of the players in each visited node or state, respectively.

3.2 Upper Confidence Bounds Applied to Trees

In MCTS, a selection strategy takes care of selecting a move for each moving player in each tree node visited during the search. The standard MCTS selection strategy is UCT (Upper Confidence bounds applied to Trees [15]). Given a state s, UCT chooses the move a^* for a player p using the UCB1 sampling strategy [1] as shown in Eq. 1.

$$a^* = \underset{a \in A(s,p)}{\operatorname{argmax}} \left\{ Q(s,a) + C \times \sqrt{\frac{\ln N(s)}{N(s,a)}} \right\}. \tag{1}$$

Here, $A(s,p)$ is the set of legal moves of player p in state s, $Q(s,a)$ is the average payoff of all simulations in which move a has been selected in state s, $N(s)$ is the number of times state s has been visited during the search and $N(s,a)$ is the number of times move a has been selected whenever state s was visited. The C constant is used to control the balance between exploitation of good moves and exploration of less visited ones.

3.3   Move-Average Sampling Technique

A successful playout strategy for MCTS is MAST [11]. The main idea behind MAST is that a move that is good in one state is likely to be good also in other states. During the search, MAST memorizes for each move a of each player p the expected payoff $Q(a,p)$ computed over all the simulations in which move a has been played so far by player p. These statistics are used to select moves during the playout. The original implementation of MAST selects moves in a state s according to the move probabilities computed using the Gibbs measure reported in Eq. 2.

$$Prob(s,a) = \frac{e^{(Q(a,p)/\tau)}}{\sum_{a' \in A(s,p)} e^{(Q(a',p)/\tau)}} \tag{2}$$

Here, a is a legal move for player p, $A(s,p)$ is the set of all the moves that are legal for player p in s, and τ is a parameter that controls the shape of the distribution. High values of τ make the distribution more uniform making it favor more exploration of the moves, while low values stretch it, making it favor more exploitation.

Later research on MAST has shown that the use of an ϵ-greedy strategy to select moves gives a significantly better performance than the use of the Gibbs measure in most of the tested games [17,23]. The ϵ-greedy strategy chooses the move with highest expected payoff $Q(a,p)$ with probability $(1-\epsilon)$ and a random move with probability ϵ.

3.4   N-Gram Selection Technique

The N-grams Selection Technique (NST) [23] is an extension of MAST. Like MAST, NST biases the playouts towards moves that have performed overall well so far. The difference is that NST also exploits information about sequences (N-grams) of consecutive moves. For each considered N-gram of moves up to a given length L, NST memorizes the average payoff obtained by all the simulations in which such N-gram of moves occurred. Note that when L is set to one, NST behaves exactly like MAST, considering statistics for single moves only.

More in detail, after each MCTS simulation, NST extracts from the simulated path all N-grams of moves up to a certain length L and uses the payoff obtained by the simulation to update their expected payoff. Duplicate N-grams in the same simulation are not detected. Therefore, the payoff of an N-gram will be updated for each of its occurrences. When extracting N-grams, NST assumes that the game is simultaneous move and each player plays a move in each turn. The N-grams of each player are updated with the payoff obtained by such player at the end of the simulation. For a game with n players, NST orders the players according to their index, $p = 1, ..., n$. For games specified in GDL, the index of a player corresponds to its order of appearance in the game rules. Assume the notation $a_{(p,t)}$ to indicate the move selected by player p during step t of the simulation. For a given player p, N-grams are extracted from a simulation with

T steps as follows:

$$\text{1-Grams} = \{\langle a_{(p,t)}\rangle | t = 1, ..., T\}$$
$$\text{2-Grams} = \{\langle a_{((p-1+n)\bmod n, t-1)}, a_{(p,t)}\rangle | t = 2, ..., T\}$$
$$\text{3-Grams} = \{\langle a_{((p-2+n)\bmod n, t-2)}, a_{((p-1+n)\bmod n, t-1)}, a_{(p,t)}\rangle | t = 3, ..., T\} \quad (3)$$
$$\vdots$$

This means that N-grams of an arbitrary length $l < L$ for player p are formed by the move of player p at a certain time step t in the simulation, preceded by the moves in the previous $l - 1$ steps of the simulation, each performed by the previous $l - 1$ players in the cyclic order, respectively.

The statistics collected for the N-grams are used by NST to select moves during the playout. The selection of a move in a state during the playout works as follows. Like for MAST, with probability ϵ a move is chosen randomly. Otherwise, with probability $(1 - \epsilon)$ the move with the highest NST value is chosen. Assume s to be the state visited during step t of the simulation, and a_p to be the move of player p for which to compute the NST value. This value is computed as follows:

$$NST(a_p) = (Q(\langle a_p\rangle) +$$
$$Q(\langle a_{((p-1+n)\bmod n, t-1)}, a_p\rangle) +$$
$$Q(\langle a_{((p-2+n)\bmod n, t-2)}, a_{((p-1+n)\bmod n, t-1)}, a_p\rangle) + \quad (4)$$
$$\vdots$$
$$Q(\langle a_{((p-L+1+n)\bmod n, t-L+1)}, ..., a_p\rangle))/L$$

When computing the NST value of a move, the expected value of the corresponding 1-gram is always considered, while the value of longer N-grams is considered only if the N-gram has been visited more than V times. This ensures that longer N-grams are used only when their value becomes more accurate.

The presented implementation of NST can also be applied to a sequential-move game by considering that players play a pass move (e.g. *noop*) when they are not on their turn. For sequential-move games, the order of players that is used by NST is the same in which the players play their turn. For some GDL games this order corresponds to the order in which roles are reported in the game rules. When this does not happen, NST re-computes the player order for each simulation by looking at which player(s) have legal moves other than *noop* in each turn. Quad is an example of such game, where a player can choose to perform more than one move before giving the control to another player. In this way, there will only be N-grams formed by meaningful moves of each player, and N-grams formed by *noop* moves only. Thus, no N-grams will mix meaningful moves with *noop* moves. This means that the statistics for the N-grams that are formed only by *noop* moves will not influence the selection of other moves.

Algorithm 1. Pseudocode for PPA

1: **procedure** PLAYOUT($s, P, W, playout$)
    Input: state s from which to start the playout, set P of all the players in the game, matrix W of move weights for each player, $playout$ that contains the states and joint moves visited so far in the current MCTS simulation from the root.
    Output: tuple $\vec{q}$ of payoffs obtained by the players in the terminal state of the current MCTS simulation.

2:    **while not** s.ISTERMINAL( ) **do**
3:        $\vec{a}^* = \langle a_1^*, ..., a_{|P|}^* \rangle \leftarrow$ empty vector of size $|P|$ ▷ empty joint move
4:        **for** $p \leftarrow 1, ..., |P|$ **do**
5:            $z \leftarrow 0.0$
6:            **for** move $a \in A(s, p)$ **do**
7:                $z \leftarrow z + exp(k \times W_{(p,a)})$
8:            $Prob \leftarrow$ empty probability distribution
9:            **for** move $a \in A(s, p)$ **do**   ▷ creation of probability distribution
10:                $Prob(a) \leftarrow \dfrac{exp(k \times W_{(p,a)})}{z}$
11:            $a_p^* \sim Prob$ ▷ sampling move from distribution
12:        add s and $\vec{a}^*$ to the $playout$
13:        $s \leftarrow$ NEXT($s, \vec{a}^*$)   ▷ advance to next state
14:        $w \leftarrow s$.GETWINNERINDEX( )
15:        **if** $w \neq$ null **then**
16:            $W \leftarrow$ ADAPT($w, W, playout$)   ▷ adapt weights if there is a winner
17:    $\vec{q} = \langle q_1, ..., q_{|P|} \rangle \leftarrow s$.GETPAYOFFS( )
18:    **return** $\vec{q}$

1: **procedure** ADAPT($w, W, playout$)
    Input: index w of the player that won the playout, matrix W of move weights for each player, $playout$ that contains all the states and joint moves visited so far in the current MCTS simulation from the root.
    Output: updated matrix of move weights for each player, V.

2:    $V \leftarrow W$ ▷ copy the weights
3:    **for** $i \leftarrow 1, ..., |playout|$ **do**
4:        $s \leftarrow playout_i$.GETVISITEDSTATE( )
5:        $\vec{a}^* \leftarrow playout_i$.GETVISITEDJOINTMOVE( )
6:        $z \leftarrow 0.0$
7:        **for** move $a \in A(s, w)$ **do**
8:            $z \leftarrow z + exp(W_{(w,a)})$
9:        **for** move $a \in A(s, w)$ **do**
10:            **if** $a = a_w^*$ **then**
11:                $V_{(w,a)} \leftarrow V_{(w,a)} + \alpha$ ▷ increment weight of visited move
12:            $V_{(w,a)} \leftarrow V_{(w,a)} - \alpha \times \dfrac{exp(W_{(w,a)})}{z}$ ▷ decrement weight of all legal moves
13:    **return** V

---

4   Simultaneous-Move Playout Policy Adaptation

The Playout Policy Adaptation (PPA) strategy [3,4] is used in MCTS to guide the search during the playout. Similarly to MAST, PPA collects information about the moves visited during the search and uses it to guide future playouts. More precisely, PPA keeps a weight for each possible move of a player, and during the playout it selects moves with a probability proportional to the exponential of their weights. At the start of the search, the weights are initialized to 0. At the end of each playout, the weights of the moves visited by the winning player are increased by a constant α. At the same time, all its legal moves in the visited states are decreased proportionally to the exponential of their weight. The α constant represents the learning rate of the algorithm. The algorithm presented in this paper memorizes the weights at the end of each turn to re-use them in the subsequent turn. This choice is motivated by previous literature, which showed that memorizing the playout policy for the PPAF algorithm is beneficial [5].

The pseudocode for the PPA playout strategy for simultaneous-move games is given in Algorithm 1. The algorithm is also applicable to sequential-move games by assuming that players play the *noop* move when it is not their turn. The procedure PLAYOUT($s, P, W, playout$) shows how one playout is performed. The procedure requires the state from which to start the playout, s, and the list of players P. Moreover, PPA uses a matrix of weights W, where each entry $W_{(p,a)}$ represents the weight for move a of player p. It also requires a list that memorizes the states and the players' joint moves visited so far during the selection phase of the current MCTS simulation (the *playout* variable in the pseudocode). During the playout, PPA selects a move a_p^* for each player p in each visited state s. To select a move for a player in s it computes a probability distribution over the legal moves proportionally to their weights. Note that the probability of a move is computed using the Gibbs measure presented in Eq. 2, where $\tau = \frac{1}{k}$. We keep the notation with k to be consistent with how previous literature introduced the PPA algorithm. Once the moves of all players have been selected, the joint move $\vec{a}^*$ is used to advance to the next state. At the end of the playout, if there is a single winner, its weights are updated by the ADAPT($w, W, playout$) procedure. Finally, a tuple of playoffs, $\vec{q}$, with an entry for each player, is returned.

The procedure ADAPT($w, W, playout$) shows how the weights of the winner are updated at the end of the playout. Other than the index of the winner w and the matrix of weights W, the procedure takes as input the *playout* variable, which now contains the complete list of states and joint moves visited during the current simulation. First of all, the matrix of weights is copied in the matrix V, so that, while V is being updated, the weights of the policy for the previous simulation can still be accessed in W. Subsequently, the algorithm iterates over all the entries in the *playout*. For each visited state s, the weight of the move played by the winner w in s is incremented by α, while the weight of all the legal moves of w in s is decreased proportionally to the exponential of their weight. This means that the sum of all the performed updates in a state is always zero.

5   PPA Enhancements

This section presents all the enhancements for PPA that are evaluated in this paper. Most of them have been previously tested for MAST [11,23,24], and are adapted in this paper to be used in combination with PPA. More precisely, Subsect. 5.1 presents the ϵ-greedy strategy for selecting moves during the playout when using the PPA strategy, and the payoff-based update of PPA weights. Next, Subsect. 5.2 introduces the after-move decay of statistics, and Subsect. 5.3 explains how PPA can be modified to consider weights for N-grams of moves instead of only for single moves. Finally, Subsect. 5.4 describes how the weights update during a simulation can be discounted depending on the playout length.

5.1 ϵ-Greedy Selection and Payoff-Based Update

As reported in Algorithm 1, when selecting a move for a player in a state, PPA uses the Gibbs measure to compute a probability distribution over the legal moves in the state and samples a move from it. As mentioned in Subsect. 3.3, previous work [23] has shown that for MAST the use of an ϵ-greedy strategy to select moves during the playout improves the performance with respect to the Gibbs measure in most of the tested games. Therefore, this paper evaluates the ϵ-greedy selection in combination with PPA. This means that during a playout, given a state s, a player p and the set of all the legal moves for p in s, $A(p,s)$, the move a_p^* in s is selected as follows:

$$a_p^* = \begin{cases} \text{random move in } A(p,s), & \text{with probability } \epsilon \\ \text{argmax}_{a \in A(p,s)} W_{(p,a)}, & \text{with probability } (1-\epsilon) \end{cases} \tag{5}$$

Here, ϵ is a predefined probability and $W_{(p,a)}$ is the PPA weight of move a of player p.

Another characteristic that distinguishes MAST from PPA is that after a simulation, MAST updates the statistics of the moves performed during the simulation for all the players, while PPA updates the weights only of the winner of the simulation. Moreover, MAST updates statistics of each player proportionally to the payoff that the player obtained in the playout. This paper proposes a modification of PPA that updates after each simulation the weights of all the players proportionally to their payoffs. Assume the payoff of the current simulations, $\vec{q} = \langle q_1, ... q_{|P|} \rangle$, to have values in the interval $[0,1]$ (note that in GDL scores have values in $[0,100]$, thus they need to be re-scaled). Given a state s in the playout and the corresponding visited joint move $\vec{a}^*$, the weights of each player p will be updated as follows:

$$W_{(p,a_p^*)} = W_{(p,a_p^*)} + \alpha \times q_p$$

$$W_{(p,a)} = W_{(p,a)} - \alpha \times q_p \times \frac{e^{W_{(p,a)}}}{\sum_{a' \in A(s,p)} e^{W_{(p,a')}}}, \forall a \in A(s,p) \tag{6}$$

5.2   After-Move Decay of Statistics

An enhancement that has been successful for MAST and NST is the decay of statistics in-between consecutive game steps [24]. MAST and NST keep the information collected during the search for a game step and reuse it in subsequent steps, so that at the start of a new search it is already known which moves are good in general. However, as the game progresses, old statistics might not be as reliable as they were before. For example, statistics collected in an early state of the game might be based on parts of the tree that are not reachable anymore at a later stage of the game. Decaying the statistics over time helps addressing this issue. Like MAST and NST, also PPA might suffer from the same issue, therefore this paper proposes to evaluate the decay of PPA weights. Three different decay methods have been proposed by Tak *et al.* [24], move decay, batch decay and simulation decay, which decay MAST statistics after each move, after a batch of simulations and after each simulation, respectively. This paper evaluates move decay, which was shown to be one of the best performing ones by Tak *et al.* [24]. Equation 7 shows how the decay with a factor of $\omega \in [0, 1]$ updates the PPA weights of a move a of player p.

$$W_{(p,a)} \leftarrow \omega \times W_{(p,a)} \tag{7}$$

5.3   N-Grams Playout Policy Adaptation

Subsection 3.4 presented the NST strategy, which has been designed by extending MAST to memorize statistics for N-grams of moves other than for single moves. A similar approach can be used to extend the PPA strategy. This paper proposes the N-gram Playout Policy Adaptation (NPPA) strategy, which is similar to PPA, but memorizes weights also for N-grams of moves.

Considering a simultaneous-move game, NPPA creates N-grams of moves in the same way described in Subsect. 3.4 for NST. In NPPA, a move for a player in a state of the playout is selected as in PPA, i.e. with Gibbs sampling. The only difference is that the weight of a move is computed as the average of the weights of the N-grams of length 1 to L. This is analogous to Eq. 4, but instead of the expected value Q, the weight W is considered for each N-gram. As NST, also NPPA considers the weights of N-grams of length greater than one only if they have been visited more than V times.

The adaptation of weights after a simulation for NPPA is slightly more complex than for NST. While NST updates statistics only of the N-Grams extracted from the simulation, NPPA updates the weights also for the N-grams that end with all the other legal moves in each visited state. More precisely, at the end of a simulation, NPPA extracts all the N-grams for the player p that won the simulation as shown in Eq. 3. Subsequently, for each of the extracted N-grams of player p it updates the weights as follows. Let $\langle a_{(t-l+1)}, ..., a_{t-1}, a_t^* \rangle$ be one of the N-grams of length l extracted from the playout for player p, where a_t^* is the move selected by player p in state s at step t of the simulation. For readability, the player index is omitted from the subscript of the moves in the N-gram, and only the simulation step is indicated. NPPA updates the following weights:

$$W_{\langle a_{(t-l+1)},...,a_{t-1},a_t^*\rangle} = W_{\langle a_{(t-l+1)},...,a_{t-1},a_t^*\rangle} + \alpha$$

$$W_{\langle a_{(t-l+1)},...,a_{t-1},a\rangle} = W_{\langle a_{(t-l+1)},...,a_{t-1},a\rangle}$$
$$- \alpha \times \frac{e^{W_{\langle a_{(t-l+1)},...,a_{t-1},a\rangle}}}{\sum_{a'\in A(s,p)} e^{W_{\langle a_{(t-l+1)},...,a_{t-1},a'\rangle}}}, \forall a \in A(s,p) \qquad (8)$$

5.4 Discount

The last enhancement evaluated in this paper consists in adding a discount to the weights update of PPA. Previous research has shown that including information about the length of the simulation when updating the statistics might improve the quality of the search [6,15]. This information is included to bias the search towards moves that favor short rather than long wins. In this paper, information about the simulation length is included in the adaptation of the weights by defining a discount factor $\gamma \in [0, 1]$. This factor decreases the magnitude of the update of the weight of a move proportionally to the distance of the move from the terminal state of the simulation. Given a state s in the simulation, and the move a^* selected by player p in s, the discounted update of the weights for p is computed as follows:

$$W_{(p,a_p^*)} = W_{(p,a_p^*)} + \alpha \times \gamma^{d(s,s_T)-1}$$

$$W_{(p,a)} = W_{(p,a)} - \alpha \times \gamma^{d(s,s_T)-1} \times \frac{e^{W_{(p,a)}}}{\sum_{a'\in A(s,p)} e^{W_{(p,a')}}}, \forall a \in A(s,p) \qquad (9)$$

Here, $d(s, s_T)$ is the distance of state s from the terminal state s_T of the simulation. This distance is computed as the number of moves that have been performed in the simulation to reach s_T from s. The fact that the distance is decremented by 1 means that the weight of the last move in the simulation that lead directly to a terminal state receives the update with full magnitude.

6 Empirical Evaluation

This section presents an empirical evaluation of the PPA playout strategy and its enhancements, comparing them with other playout strategies: random, MAST and NST. Multiple series of experiments have been performed. First, Subsect. 6.1 describes the general setup of the experiments. Next, Subsects. 6.2, 6.3, 6.4 and 6.5 present the evaluation of the basic version of PPA, the ϵ-greedy selection of moves and payoff-based adaptation of the policy, the use of a decay factor and the use of N-grams, respectively. Subsection 6.6 evaluates the performance of MAST, NST and the best version of PPA and NPPA for different simulation budgets. Finally, Subsection 6.8 presents the results of the application of the depth-based discount to the payoffs.

6.1 Setup

The strategies evaluated in this paper have been implemented in the framework provided by the open source GGP-Base project [20], as part of an MCTS agent that follows the setup described by Sironi [21]. Experiments are carried out on a set of 14 heterogeneous games taken from the GGP Base repository [19]: 3D Tic Tac Toe, Breakthrough, Knightthrough, Chinook, Chinese Checkers with 3 players, Checkers, Connect 5, Quad (the version played on a 7 × 7 board), Sheep and Wolf, Tic-Tac-Chess-Checkers-Four (TTCC4) with 2 and 3 players, Connect 4, Pentago and Reversi.

Five basic playout strategies are used in the experiments: random, MAST, NST, PPA and NPPA. All of them are combined with the UCT selection strategy with $C = 0.7$. The fixed settings of these strategies are summarized below. For PPA and NPPA some settings vary depending on the experiment, thus, their values will be specified in the corresponding subsections. MAST and NST use an ϵ-greedy strategy to select moves with $\epsilon = 0.4$, and decay statistics after each move with a factor of 0.2 (i.e. 20% of the statistics is reused in the subsequent turn). NST uses N-grams up to length $L = 3$, and considers them only if visited at least $V = 7$ times. All the mentioned parameter values are taken from previous literature, where they were shown to perform well [21,23]. PPA and NPPA set $k = 1$, as in previous publications on PPA [3–5], when using the Gibbs measure to select moves. When using the ϵ-greedy selection, they set $\epsilon = 0.4$, like for MAST. For NPPA, the same values of $L = 3$ and $V = 7$ that are set for NST are used.

For each performed experiment, two strategies at a time are matched against each other. For each game, all possible assignments of strategies to the roles are considered, except the two configurations that assign the same strategy to all roles. All configurations are run the same number of times until at least 500 games have been played. Each match runs with a budget of 1000 simulations per turn, except for the experiments in Subsect. 6.6, that test different budgets. Experimental results always report the average win percentage of one of the two involved strategies with a 95% confidence interval. The average win percentage for a strategy on a game is computed looking at the score. The strategy that achieves the highest score is the winner and gets 1 point, if both achieve the highest score they receive 0.5 points each, and it is considered a tie. In each table, bold results represent the highest win percentage for each game in the considered part of the table.

6.2 Basic PPA

This series of experiments evaluates the basic version of the PPA playout strategy presented in Algorithm 1. PPA is compared with MAST by matching both of them against a random playout strategy, and by matching them directly against each other. Two values for the learning rate α are evaluated, 1 and 0.32. The first one is the value used by the first proposed version of PPA [3]. However,

Table 1. Win% of different configurations of PPA against a random playout strategy and against MAST with a budget of 1000 simulations.

Game	vs random playout strategy			vs MAST	
	PPA_1	$PPA_{0.32}$	MAST	PPA_1	$PPA_{0.32}$
3DTicTacToe	78.2(±3.62)	**89.4**(±2.70)	88.8(±2.77)	38.1(±4.26)	**52.7**(±4.37)
Breakthrough	66.0(±4.16)	**66.4**(±4.14)	62.2(±4.25)	62.2(±4.25)	**63.0**(±4.24)
Knightthrough	51.6(±4.38)	53.4(±4.38)	**85.2**(±3.12)	12.2(±2.87)	11.0(±2.75)
Chinook	64.0(±3.28)	65.7(±3.26)	**69.3**(±3.30)	45.9(±3.64)	**50.8**(±3.56)
C.Checkers 3P	64.7(±4.18)	71.4(±3.95)	**74.0**(±3.83)	46.6(±4.36)	**52.4**(±4.36)
Checkers	60.0(±4.00)	68.1(±3.78)	**75.3**(±3.44)	35.6(±3.91)	**37.1**(±3.92)
Connect 5	63.8(±4.17)	75.9(±3.69)	**80.7**(±3.27)	33.2(±3.55)	**49.5**(±3.58)
Quad	73.9(±3.85)	81.7(±3.36)	**82.1**(±3.35)	36.7(±4.18)	**45.7**(±4.33)
SheepAndWolf	59.2(±4.31)	**61.4**(±4.27)	46.0(±4.37)	69.2(±4.05)	**69.4**(±4.04)
TTCC4 2P	75.4(±3.78)	**76.8**(±3.69)	76.0(±3.74)	50.8(±4.37)	**58.1**(±4.32)
TTCC4 3P	59.3(±4.28)	**61.5**(±4.23)	59.2(±4.27)	**53.9**(±4.30)	49.0(±4.34)
Connect 4	24.8(±3.71)	**36.3**(±4.12)	27.3(±3.80)	48.2(±4.30)	**55.9**(±4.27)
Pentago	55.0(±4.32)	62.9(±4.20)	**72.1**(±3.89)	27.1(±3.86)	**39.2**(±4.23)
Reversi	55.3(±4.30)	62.7(±4.19)	**82.8**(±3.26)	21.9(±3.58)	**26.0**(±3.72)
Avg Win%	60.8(±1.12)	66.7(±1.08)	**70.1**(±1.05)	41.6(±1.12)	**47.1**(±1.13)

subsequent research has shown the value 0.32 to perform better for the PPAF variant of the algorithm [5], therefore it is considered here as well.

Table 1 reports the results, indicating the two PPA variants with PPA_1 and $PPA_{0.32}$, respectively. First of all, we can see that both PPA variants and MAST are better than the random playout strategy. Only in Connect 4 they all perform significantly worse than random playouts. When comparing PPA_1 and $PPA_{0.32}$, it is visible that the latter has a significantly better overall performance than the former, both against the random playout strategy and against MAST. Moreover, looking at individual games, the performance of $PPA_{0.32}$ is never significantly inferior to the performance of PPA_1, being significantly higher in a few of the games. In 3D Tic Tac Toe, Connect 5 and Quad the win rate of $PPA_{0.32}$ is higher both when the opponent uses random playouts and when it uses MAST. Moreover, against the random opponent $PPA_{0.32}$ seems better than PPA_1 in Checkers and Connect 4, while against MAST it seems better in Pentago.

When comparing the performance of PPA with the one of MAST, it is clear that the latter has a significantly higher overall win rate. However, for $PPA_{0.32}$ the difference with MAST is smaller, especially when PPA is matched against MAST directly. It is also interesting to look at the performance of MAST and PPA on individual games. MAST seems a particularly suitable strategy for Knightthrough, for which it achieves a much higher win rate than PPA. Also for Pentago, Reversi and Checkers MAST seems to perform better than PPA. On the contrary, PPA seems to be particularly suitable for Sheep and Wolf and, when matched against MAST directly, for Breakthrough as well. Finally, when

Table 2. Win% of different configurations of PPA against MAST with a budget of 1000 simulations and $\alpha = 0.32$.

Game	$PPA_{0.32}$	$PPA_{P0.32}$	$PPA_{\epsilon 0.32}$	$PPA_{\epsilon P0.32}$
3DTicTacToe	52.7($\pm$4.37)	**61.0**($\pm$4.26)	27.5($\pm$3.90)	34.1($\pm$4.15)
Breakthrough	63.0($\pm$4.24)	**65.2**($\pm$4.18)	58.2($\pm$4.33)	59.4($\pm$4.31)
Knightthrough	11.0($\pm$2.75)	12.2($\pm$2.87)	**38.2**($\pm$4.26)	37.0($\pm$4.24)
Chinook	50.8($\pm$3.56)	**53.7**($\pm$3.61)	37.0($\pm$3.41)	34.8($\pm$3.50)
C.Checkers 3P	**52.4**($\pm$4.36)	41.5($\pm$4.31)	46.2($\pm$4.36)	26.2($\pm$3.84)
Checkers	37.1($\pm$3.92)	**42.6**($\pm$3.99)	26.4($\pm$3.59)	25.7($\pm$3.48)
Connect 5	49.5($\pm$3.58)	**57.3**($\pm$3.81)	39.3($\pm$3.74)	32.9($\pm$3.84)
Quad	45.7($\pm$4.33)	**47.2**($\pm$4.34)	29.7($\pm$3.97)	33.5($\pm$4.12)
SheepAndWolf	69.4($\pm$4.04)	**71.2**($\pm$3.97)	61.6($\pm$4.27)	63.4($\pm$4.23)
TTCC4 2P	**58.1**($\pm$4.32)	56.0($\pm$4.33)	50.4($\pm$4.36)	42.7($\pm$4.33)
TTCC4 3P	49.0($\pm$4.34)	**53.1**($\pm$4.33)	47.6($\pm$4.34)	49.2($\pm$4.33)
Connect 4	**55.9**($\pm$4.27)	51.3($\pm$4.31)	42.1($\pm$4.22)	44.7($\pm$4.28)
Pentago	39.2($\pm$4.23)	**41.3**($\pm$4.30)	35.0($\pm$4.13)	34.6($\pm$4.15)
Reversi	26.0($\pm$3.72)	29.7($\pm$3.93)	**38.0**($\pm$4.20)	37.6($\pm$4.16)
Avg Win%	47.1($\pm$1.13)	**48.8**($\pm$1.14)	41.2($\pm$1.12)	39.7($\pm$1.12)

$\alpha = 0.32$, PPA performs significantly better against MAST in TTCC4 with 2 players and Connect 4.

The results presented in Table 1 for Breakthrough and Knightthrough can also be compared with results obtained for these games in previous work on PPA by Cazenave [4]. First of all, as shown by previous literature, Table 1 confirms that PPA and MAST perform significantly better than random playouts on these two games. Moreover, for Knightthrough, MAST clearly performs better than PPA, as it was also shown in previous work, where MAST achieved a win rate of 78.2%, while PPA, for $\alpha = 0.32$, only reached 70.4%. Note that in Table 1 the difference in performance seems to be even higher. This could be due to the fact that the experiments in this paper have some different settings than the ones used by Cazenave. For example, MAST is using an ϵ-greedy strategy to select moves, which has been shown to perform generally better than the one based on the Gibbs measure that is used by Cazenave. Moreover, the search budget used by Cazenave is higher than the 1000 simulations used in Table 1. For Breakthrough, Table 1 seems to suggest that PPA performs slightly better than MAST, while in previous work it seemed to be the opposite, with MAST reaching a win rate of 63.4% and PPA with $\alpha = 0.32$ only of 59.2. However, the difference between MAST and PPA does not seem to be particularly high in either case. The difference between this and previous work might be due to the difference in the experimental setup.

6.3 ϵ-Greedy Selection and Payoff-Based Update

This series of experiments looks at the impact of two modifications of PPA inspired by MAST: the use of an ϵ-greedy strategy to select moves and the update of statistics based on the payoffs obtained by the players during a simulation. Given that $PPA_{0.32}$ performed best in the previous series of experiments, it is used as a baseline for this series of experiments to extend upon. The considered PPA variants are $PPA_{\epsilon 0.32}$ (uses ϵ-greedy selection and updates weights only of the playout winner), $PPA_{P0.32}$ (uses Gibbs-based move selection and payoff-based weight update), and $PPA_{\epsilon P0.32}$ (uses ϵ-greedy selection and payoff-based weight update). All variants of PPA are matched only against MAST, because it is a more competitive agent than the one that uses random playouts and it is more interesting to compare against.

Table 2 reports the win rate of the variants of PPA. Looking at the overall win rate, it seems that the ϵ-greedy selection does not have on PPA the same beneficial impact that it has been shown to have on MAST with respect to the use of the Gibbs measure [23]. Basing the selection on the Gibbs measure seems to perform best for PPA. Moreover, the payoff-based update of the policy does not seem to have a significant overall impact with respect to the update based only on the winner. Finally, all considered variants of PPA are still performing worse than MAST, although the performance of PPA that uses the Gibbs measure is rather close.

More interesting results emerge when looking at individual games. First of all, none of the considered PPA variants performs better than the others on all games, but the performance is game-dependent. Furthermore, even, if in most games it significantly reduces the performance, the ϵ-greedy selection seems to significantly increase the win rate of PPA in Knightthrough and Reversi. This suggest that the use of the ϵ-greedy strategy might be one of the reasons why in the previous series of experiments the MAST strategy achieved a high win rate on these games. Thus, these games might benefit from the use of such selection of moves. Finally, it can be noted that there are still some games for which all variants of PPA perform significantly better than MAST, namely Breakthrough and Sheep and Wolf. Moreover, $PPA_{P0.32}$ is the variant that performs better than MAST in more games than the other variants. Other than in the already mentioned games, it is significantly better than MAST also in 3D Tic Tac Toe, Connect 5 and Chinook. For this reason, this is the variant that will be further enhanced and evaluated in the next series of experiments. In addition, this variant might have the advantage of speeding up the learning process because after each simulation it adapts the policy of each player, and not only of the winner.

6.4   After-Move Decay of Statistics

This series of experiments evaluates different values for the after-move decay of PPA statistics. The PPA strategy that performed best in the previous series of experiments, $PPA_{P0.32}$ is considered. Figure 1 shows how the average win rate of $PPA_{P0.32}$ against MAST changes for different decay factors. Once again, we

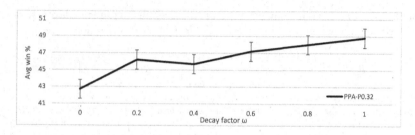

Fig. 1. Win% of PPA$_{P0.32}$ with different values for the statistics decay ω against MAST with a budget of 1000 simulations.

can see that the performance of PPA$_{P0.32}$ does not surpass the one of MAST. Focusing on PPA$_{P0.32}$, it seems that higher values of ω perform overall best. Thus, it seems better for PPA$_{P0.32}$ to reuse most, if not all, the collected statistics for subsequent game steps. This means that the after-move decay of statistics that works well for MAST, might not have the same impact on PPA. For this reason, no decay is applied to PPA statistics in further experiments.

Results for individual games are not reported, because they simply seem to confirm that high values for ω should be preferred. Result worth mentioning are the ones obtained for Knightthrough. For this game, results showed that high values of ω are detrimental, while low values improve the performance substantially. For example, no decay of statistics (i.e. $\omega = 1$) reaches a win rate of only 12.2($\pm$2.87), while a large decay of statistics (in this case $\omega = 0.2$) reaches the win rate of 48.6($\pm$4.39). Note that this might, once again, be part of the reason why in the previous series of experiments MAST performed much better than PPA on this game. The implementation of MAST used in this paper is also performing after-move decay of statistics with a factor 0.2.

6.5   N-Grams Playout Policy Adaptation

This series of experiments evaluates the NPPA playout strategy, which behaves like PPA, but in addition uses weights for N-grams of moves. Two settings for the strategy have been tested. The first one corresponds to the initial setting of PPA evaluated in Subsect. 6.2 and taken from previous literature [3,5]. Thus, NPPA with move selection based on Gibbs, policy update only for the winner and $\alpha = 1$ (NPPA$_1$). The second setting corresponds to the best settings found in previous experiments for PPA. Thus, with move selection based on Gibbs, payoff-based policy update for all players, and $\alpha = 0.32$ (NPPA$_{P0.32}$). These two variants of NPPA are compared with NST by matching them all against random playouts and by matching them against each other. Moreover, NPPA is also matched directly against the best version of PPA tested so far, i.e. PPA$_{P0.32}$.

Looking at results in Table 3, it is evident that both NPPA and NST perform overall significantly better then random playouts. In addition, none of them performs significantly worse than random playouts on any of the games, except

Table 3. Win% of different configurations of NPPA against a random playout strategy, against NST, and against PPA$_{P0.32}$ with a budget of 1000 simulations.

Game	vs random playout strategy			vs NST		vs PPA$_{P0.32}$	
	NPPA$_1$	NPPA$_{P0.32}$	NST	NPPA$_1$	NPPA$_{P0.32}$	NPPA$_1$	NPPA$_{P0.32}$
3DTicTacToe	71.6(±3.96)	51.4(±4.39)	**93.4**(±2.18)	20.6(±3.55)	8.4(±2.43)	**24.8**(±3.79)	9.4(±2.56)
Breakthrough	**73.6**(±3.87)	52.6(±4.38)	68.8(±4.07)	**58.4**(±4.32)	37.0(±4.24)	**57.6**(±4.34)	38.4(±4.27)
Knightthrough	74.0(±3.85)	57.2(±4.34)	**89.8**(±2.66)	28.4(±3.96)	17.6(±3.34)	**76.4**(±3.73)	52.0(±4.38)
Chinook	62.2(±3.31)	53.2(±3.50)	**67.9**(±3.45)	47.8(±3.70)	33.4(±3.36)	**47.2**(±3.40)	36.3(±3.41)
C.Checkers 3P	**71.4**(±3.95)	53.2(±4.36)	68.5(±4.06)	**53.0**(±4.36)	30.8(±4.03)	**61.9**(±4.24)	35.7(±4.19)
Checkers	72.3(±3.66)	48.4(±4.13)	**73.8**(±3.54)	**48.6**(±4.04)	24.7(±3.44)	**58.8**(±3.92)	34.8(±3.81)
Connect 5	61.6(±4.20)	51.2(±4.39)	**84.0**(±2.99)	30.5(±3.47)	18.2(±3.10)	**36.5**(±3.97)	21.8(±3.58)
Quad	75.7(±3.76)	51.8(±4.38)	**87.7**(±2.85)	38.3(±4.23)	16.6(±3.25)	**42.8**(±4.31)	19.9(±3.50)
SheepAndWolf	65.2(±4.18)	50.6(±4.39)	49.0(±4.39)	**72.2**(±3.93)	52.0(±4.38)	**48.2**(±4.38)	39.0(±4.28)
TTCC4 2P	71.2(±3.95)	53.8(±4.36)	**72.1**(±3.92)	46.0(±4.37)	25.7(±3.83)	**45.2**(±4.35)	25.7(±3.83)
TTCC4 3P	**63.5**(±4.16)	52.2(±4.36)	59.8(±4.26)	**55.3**(±4.31)	41.2(±4.25)	**50.7**(±4.35)	39.3(±4.23)
Connect 4	44.5(±4.29)	**48.7**(±4.27)	47.2(±4.28)	42.6(±4.20)	**51.8**(±4.27)	61.1(±4.21)	**67.7**(±3.99)
Pentago	59.6(±4.26)	54.6(±4.33)	**70.2**(±3.96)	**42.7**(±4.29)	30.5(±3.98)	**46.5**(±4.35)	40.5(±4.25)
Reversi	**84.7**(±3.09)	55.5(±4.30)	80.4(±3.40)	**55.3**(±4.31)	18.4(±3.32)	**72.5**(±3.84)	40.6(±4.25)
Avg Win%	67.9(±1.07)	52.5(±1.14)	**72.3**(±1.02)	**45.7**(±1.13)	29.0(±1.03)	**52.2**(±1.14)	35.8(±1.10)

for NPPA$_1$ in Connect 4. This seems to confirm results in Table 1, where random playouts seemed to work better than more informed playouts for this game. Comparing NPPA with NST, both variants of the former are performing significantly worse than the latter, especially NPPA$_{P0.32}$. One of the reasons for this could be that NPPA is updating statistics of N-grams more often than NST. NPPA does not update only statistics of visited N-grams of moves, but also of N-grams that end with all the other legal moves in each visited state. This means that at the start of the search, statistics are updated more often using results of less reliable playouts, and it might take longer for NPPA to converge to optimal values for the policy. This would also explain why $\alpha = 1$ seems to have a better performance than $\alpha = 0.32$. When updating weights, a higher learning rate enables NPPA to give more importance to the result of new and possibly more reliable simulations. There are still a few games where NPPA$_1$ performs significantly better than NST, namely Breakthrough, TTCC4 with 3 players, Reversi and Sheep and Wolf. In the latter, also NPPA$_{P0.32}$ shows a good performance.

Results also show that NPPA$_1$ is significantly better than NPPA$_{P0.32}$ in almost all the games, and for each of the tree different opponents they are matched against (i.e. random, NST, PPA$_{P0.32}$). Therefore, the setting that achieved the highest results for PPA do not seem to work as well for NPPA. When matched against PPA$_{P0.32}$, NPPA$_{P0.32}$ has a significantly lower win rate in all games except Connect 4. On the contrary, NPPA$_1$ has a higher overall win rate, and performs significantly worse only on four games (3D Tic Tac Toe, Connect 5, Quad and TTCC4 with 2 players). In Knightthrough, NPPA$_1$ achieves a much higher win rate than PPA$_{P0.32}$. This is in part likely due to the settings of PPA$_{P0.32}$ being sub-optimal for this game, but might also indicate that collecting statistics for N-grams of moves improves the selection of moves during the playout for this game. Finally, the performance of NPPA can be compared with

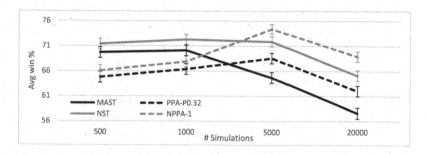

Fig. 2. Win% of MAST, PPA$_{P0.32}$, NST and NPPA$_1$ against a random playout strategy for different numbers of simulations per turn.

the one of PPA against random playouts, that is reported in Table 2. Considering the best version of NPPA (i.e. NPPA$_1$), it is visible that its overall performance against random playouts is significantly higher than the one of PPA$_1$, and at least equal to the one of PPA$_{0.32}$. Moreover, it is significantly better than PPA$_{0.32}$ in Knightthrough and Reversi, for which the general PPA strategy does not seem to perform well against MAST.

6.6 Simulation Budget Analysis

This series of experiments analyzes the performance of MAST, PPA$_{P0.32}$, NST and NPPA$_1$ with different numbers of simulations per turn. The versions of PPA and NPPA considered for the experiments are the ones that seemed to perform best in previous experiments. PPA and NPPA are updating the statistics of more moves per simulation than MAST and NST, respectively. This means that it might require more time for this statistics to be reliable. This motivates the interest in analyzing how these strategies are influenced by the search budget.

Figure 2 shows how the win rate over all the games of the considered strategies changes against the random playout strategy depending on the budget. For most of the considered games, the individual plots look very similar. The plot shows that the performance of all algorithms eventually decreases with more simulations. Moreover, we can see how higher budgets benefit PPA$_{P0.32}$ and NPPA$_1$ more than MAST and NST. According to the law of diminishing returns [14], the performance gain of a search algorithm decreases with the increase of search effort. In this case it looks like the decrease in gain for random playouts is the slowest, thus we notice a decrease of performance over time for all other strategies. However, for PPA$_{P0.32}$ and NPPA$_1$, the decrease in gain seems slower than for NST and MAST. Thus, PPA$_{P0.32}$ and NPPA$_1$ are able to surpass the performance of MAST and NST, respectively, when the budget is increased to 5000 simulations and higher. It seems that PPA$_{P0.32}$ and NPPA$_1$ require a higher number of simulations than MAST and NST in order to converge to a good policy, but then they can surpass the performance of MAST and NST, respectively.

The plot also confirms that the use of N-grams can significantly improve the performance of both MAST and PPA. It also shows that the impact of N-grams

Table 4. Time (ms) that each playout strategy takes to perform 1000 simulations.

Game	Random	MAST	PPA$_{P0.32}$	NST	NPPA$_1$
3DTicTacToe	352	196	1357	268	2362
Breakthrough	683	226	604	275	1374
Knightthrough	492	166	470	190	633
Chinook	485	281	1568	324	1882
C.Checkers 3P	191	171	539	191	731
Checkers	1974	1923	3545	1661	4164
Connect 5	822	404	2061	520	3100
Quad	435	300	1248	381	1691
SheepAndWolf	471	379	718	406	1008
TTCC4 2P	1069	651	773	745	1465
TTCC4 3P	787	599	650	619	981
Connect 4	102	89	245	98	357
Pentago	394	290	1212	357	1668
Reversi	3562	3493	3521	3433	4525

on the search is larger for higher budgets. This is not surprising, as it takes some time for the algorithms to collect enough samples for the N-gram statistics to be reliable and positively influence the search.

6.7   Time Analysis

An important aspect to consider when comparing MAST, PPA, NST and NPPA is the computational cost of these strategies. As opposed to random playouts, they invest part of the computational time into memorizing statistics and computing values on which their move-selection is based. Table 4 shows the time in milliseconds that each strategy requires to perform 1000 simulations per turn. These values have been obtained averaging the median time per turn over 500 runs of each game. For consistency, these experiments have been all performed on the same machine, a Linux server consisting of 64 AMD Opteron 6274 2.2-GHz cores. Although all strategies spend time computing statistics, it is immediately clear that both MAST and NST are able to make up for this loss. Both of them never take more time than random playouts to perform 1000 simulations. This indicates that, by biasing the search towards promising moves, they are probably visiting shorter paths that lead to a faster conclusion of the simulated game. Unfortunately, the same effect is not observed for PPA$_{P0.32}$ and NPPA$_1$, except possibly for PPA$_{P0.32}$ in Breakthrough, Knightthrough and TTCC4 with 2 and 3 players, where the search time is lower than the one of random playouts. It could be that PPA$_{P0.32}$ is visiting shorter playouts as well, but the time saved is not sufficient to make up for the extra computation involved. These results are not surprising, given the fact that PPA$_{P0.32}$ and NPPA$_1$ are updating more

Table 5. Average win rate over all games of each playout strategy against the random playout strategy, with time budgets of 1 and 5 s.

	MAST	$PPA_{P0.32}$	NST	$NPPA_1$
1 s	70.1($\pm$1.04)	50.9($\pm$1.14)	73.7($\pm$0.99)	48.0($\pm$1.14)
5 s	60.4($\pm$1.09)	52.8($\pm$1.12)	67.3($\pm$1.05)	53.8($\pm$1.12)

statistics per simulation than MAST and NST, and indicate a possible limitation of the two former strategies. Although more accurate in the long run, $PPA_{P0.32}$ and $NPPA_1$ might be less suitable than MAST and NST for domains with short time settings, as they might not be able to reach a sufficient number of simulations for their statistics to become accurate and make a positive difference in performance with respect to MAST and NST, respectively. This seems to be confirmed by how the performance of all the strategies changes when they are compared for different time budgets instead of simulation budgets. Table 5 shows the average win rate of the strategies against random playouts over all the considered games. Time budgets of 1 and 5 s are considered. Results seem to indicate that the advantage of MAST and NST over random playouts decreases with a higher time budget, while for PPA and NPPA it seems to increase. This means that with more time $PPA_{P0.32}$ and $NPPA_1$ might have an advantage over MAST and NST, respectively. For a couple of games, this already seems to happen with a time budget of 5 s. For example, in TTCC4 with 3 players, the win rate of MAST against random playouts changes from 58.7($\pm$4.25) for a budget of 1 s to 56.2($\pm$4.25) for a budget of 5 s, while for $PPA_{P0.32}$ it increases from 54.8($\pm$4.28) to 66.0($\pm$4.02). Moreover, the win rate of NST changes from 60.4($\pm$4.20) to 58.8($\pm$4.17), while for $NPPA_1$ it increases from 57.7($\pm$4.28) to 72.6($\pm$3.78). Another example is Sheep and Wolf, for which $PPA_{P0.32}$ performs better than MAST against random playouts both for 1 and 5 s. Moreover, for 5 s the performance difference between the two strategies seems to become higher. For 1 s, MAST has a win rate of 45.2($\pm$4.37) and $PPA_{P0.32}$ has a win rate of 55.4($\pm$4.36), while for 5 s MAST has a win rate of 42.4($\pm$4.34) and $PPA_{P0.32}$ has a win rate of 61.2($\pm$4.28).

6.8   Discount

In the final series of experiment, the effect of discounting the payoffs depending on the depth reached by the playout is evaluated on all the strategies considered in this paper, MAST, PPA, NST and NPPA. Once again, we consider the best versions of PPA and NPPA, i.e. $PPA_{P0.32}$ and $NPPA_1$. All the strategies are matched against the random playout strategy. For MAST and NST the discount is applied directly to the payoff that is being backpropagated from leaf to root. Thus, for each backpropagation step the payoff gets further discounted with the factor γ. Three discount factors are considered, 1 (i.e. no discount), 0.99 and 0.95. High values of γ are used to avoid the payoff to become too small close to the root for long playouts.

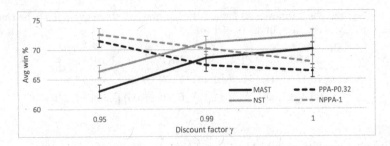

Fig. 3. Win% of MAST, PPA, NST and NPPA with different discount factors against a random playout strategy with a budget of 1000 simulations.

The results of this series of experiments are reported in Fig. 3 over all the tested games. The plot shows a performance increase of PPA$_{P0.32}$ and NPPA$_1$ with the decrease of the discount factor, while the performance of MAST an NST decreases. The discount seems to be beneficial only for PPA and NPPA. With $\gamma = 0.95$ they are able to reach the same performance of MAST and NST with no discount. Moreover, they can achieve a significantly higher performance than MAST and NST that use a discount of 0.95. Most of the plots for individual games follow a very similar trend to the one shown in Fig. 3. An exception is NPPA$_1$ in Checkers, that experiences a decrease in performance, going from 72.3($\pm$3.66) with $\gamma = 1$, to 63.6($\pm$3.92) with $\gamma = 0.95$. Another exception is NST in TTCC4 with two players, that experiences an increase in performance, going from 72.1($\pm$3.92) with $\gamma = 1$, to 81.9($\pm$3.36) with $\gamma = 0.95$.

7   Conclusion and Future Work

This paper investigated Playout Policy Adaptation (PPA) as a domain independent playout strategy for MCTS that can be applied to General Game Playing. First, its formulation for simultaneous move games has been presented. Subsequently, five different enhancements for this strategy have been introduced and evaluated: the ϵ-greedy selection of moves, the update of move statistics of all players proportionally to their payoffs, the decay of statistics in-between subsequent game steps, the use of statistics collected for N-grams of moves (which defines the new NPPA strategy), and a depth-based discount of payoffs during backpropagation. Most of these enhancements had already been successfully applied to the similar MAST playout strategy.

First of all, experiments show that PPA and NPPA are suitable to be applied to a wide variety of games, as they achieve competitive results not only in two-player, sequential-move games, but also in multi-player games (e.g. Chinese Checkers and TTCC4 with 3 players) and simultaneous-move games (e.g. Chinook). It may be concluded that both PPA and NPPA with appropriate settings perform significantly better than random playouts.

Despite sharing some similarities, MAST and PPA do not seem to show the same performance under similar settings or similar circumstances. First of all, considering a budget of 1000 simulations, not all enhancements that have been shown to work well for MAST are helpful for PPA as well, and not all enhancements that work well for PPA might have the same effect on MAST. Given the presented results, it may be concluded that the use of an ϵ-greedy strategy and of the after-move decay of statistics, especially with a low decay factor, is detrimental for the performance of PPA. On the contrary, it may be concluded that using a payoff-based update of the policy or collecting statistics for N-grams is generally not detrimental and seems to improve the performance in a few games. For N-grams, however, it is important to note that the improvement of the performance might depend also on how other parameters are set. The setup of NPPA that performed best is, indeed, different than the best set-up found for PPA. The depth-based discount of payoffs during backpropagation has also been shown to have a different influence on the considered strategies. It improves the overall performance of PPA and its NPPA variant, but is not as helpful for MAST and NST, even causing a decrease in their performance. Thus, its use is recommended only for PPA and NPPA.

Another characteristic that distinguishes PPA and MAST is that the former performs better with a higher budget. This is likely because PPA needs more time to collect accurate statistics, but then it might be able to converge to a better policy. The same can be seen when comparing NPPA and NST. Thus, it may be concluded that PPA and NPPA might be more suitable than MAST and NST when there is a higher budget available. Finally, it can be concluded that, for a low search budget, PPA and NPPA seem to have an equivalent performance to MAST and NST, respectively. For a budget of 1000 simulations, experimental results have shown that the best versions of PPA and NPPA (i.e. $PPA_{GP0.32}$ and $NPPA_{G1}$), when combined with a discount $\gamma = 0.95$ have a similar performance to the state-of-the-art implementation of MAST and NST.

It is important to keep in mind that this work did not explore all combinations of parameter values and enhancements. Therefore, some good configurations for all the considered strategies might have been missed. For example, more values could be evaluated for the discount factor γ. For PPA and NPPA, the α constant could be better tuned. Moreover, for some of the enhancements applied to PPA, parameters were set to values that performed well on MAST (e.g. the value of ϵ or the values of L and V when using N-grams), but different values might be better for PPA. Finally, various combinations of the enhancements were evaluated for PPA, while the starting configurations considered for NPPA where based on the performance observed on PPA. Different combinations that were not considered might work better for NPPA.

It is also worth mentioning that the results highlight that no single strategy and no single setting for a given strategy is able to perform best on all games. This shows that, in a domain like GGP, the search has to be adapted to each game online, instead of using fixed strategies with fixed control-parameters. This could be a promising direction for future work.

Acknowledgments. This work was supported in part by the French government under management of Agence Nationale de la Recherche as part of the "Investissements d'avenir" program, reference ANR19-P3IA-0001 (PRAIRIE 3IA Institute).

References

1. Auer, P., Cesa-Bianchi, N., Fischer, P.: Finite-time analysis of the multiarmed bandit problem. Mach. Learn. **47**(2–3), 235–256 (2002). https://doi.org/10.1023/A:1013689704352
2. Browne, C.B., et al.: A survey of Monte Carlo tree search methods. IEEE Trans. Comput. Intell. AI Games **4**(1), 1–43 (2012)
3. Cazenave, T.: Playout policy adaptation for games. In: Plaat, A., van den Herik, J., Kosters, W. (eds.) ACG 2015. LNCS, vol. 9525, pp. 20–28. Springer, Cham (2015). https://doi.org/10.1007/978-3-319-27992-3_3
4. Cazenave, T.: Playout policy adaptation with move features. Theoret. Comput. Sci. **644**, 43–52 (2016)
5. Cazenave, T., Diemert, E.: Memorizing the playout policy. In: Cazenave, T., Winands, M.H.M., Saffidine, A. (eds.) CGW 2017. CCIS, vol. 818, pp. 96–107. Springer, Cham (2018). https://doi.org/10.1007/978-3-319-75931-9_7
6. Cazenave, T., Saffidine, A., Schofield, M., Thielscher, M.: Nested Monte Carlo search for two-player games. In: Thirtieth AAAI Conference on Artificial Intelligence, pp. 687–693 (2016)
7. Cazenave, T., Teytaud, F.: Application of the nested rollout policy adaptation algorithm to the traveling salesman problem with time windows. In: Hamadi, Y., Schoenauer, M. (eds.) LION 2012. LNCS, vol. 7219, pp. 42–54. Springer, Heidelberg (2012). https://doi.org/10.1007/978-3-642-34413-8_4
8. Coulom, R.: Efficient selectivity and backup operators in Monte-Carlo tree search. In: van den Herik, H.J., Ciancarini, P., Donkers, H.H.L.M. (eds.) CG 2006. LNCS, vol. 4630, pp. 72–83. Springer, Heidelberg (2007). https://doi.org/10.1007/978-3-540-75538-8_7
9. Edelkamp, S., Gath, M., Cazenave, T., Teytaud, F.: Algorithm and knowledge engineering for the TSPTW problem. In: 2013 IEEE Symposium on Computational Intelligence in Scheduling (CISched), pp. 44–51. IEEE (2013)
10. Edelkamp, S., Gath, M., Greulich, C., Humann, M., Herzog, O., Lawo, M.: Monte-Carlo tree search for logistics. In: Clausen, U., Friedrich, H., Thaller, C., Geiger, C. (eds.) Commercial Transport. LNL, pp. 427–440. Springer, Cham (2016). https://doi.org/10.1007/978-3-319-21266-1_28
11. Finnsson, H., Björnsson, Y.: Learning simulation control in General Game-Playing agents. In: Twenty-Fourth AAAI Conference on Artificial Intelligence (AAAI), pp. 954–959. AAAI Press (2010)
12. Genesereth, M., Love, N., Pell, B.: General game playing: overview of the AAAI competition. AI Mag. **26**(2), 62–72 (2005)
13. Genesereth, M., Thielscher, M.: General Game Playing. Synthesis Lectures on Artificial Intelligence and Machine Learning, vol. 8. Morgan & Claypool Publishers, San Rafael (2014)
14. Heinz, E.A.: Self-play, deep search and diminishing returns. ICGA J. **24**(2), 75–79 (2001)
15. Kocsis, L., Szepesvári, C.: Bandit based Monte-Carlo planning. In: Fürnkranz, J., Scheffer, T., Spiliopoulou, M. (eds.) ECML 2006. LNCS (LNAI), vol. 4212, pp. 282–293. Springer, Heidelberg (2006). https://doi.org/10.1007/11871842_29
16. Love, N., Hinrichs, T., Haley, D., Schkufza, E., Genesereth, M.: General game playing: game description language specification. Technical report, Stanford Logic Group (2008)

17. Powley, E.J., Whitehouse, D., Cowling, P.I.: Bandits all the way down: UCB1 as a simulation policy in Monte Carlo tree search. In: 2013 IEEE Conference on Computational Intelligence in Games (CIG), pp. 81–88. IEEE (2013)

18. Rosin, C.D.: Nested rollout policy adaptation for Monte Carlo tree search. In: IJCAI, pp. 649–654 (2011)

19. Schreiber, S.: Games - base repository (2016). http://games.ggp.org/base/

20. Schreiber, S., Landau, A.: The General Game Playing base package (2016). https://github.com/ggp-org/ggp-base

21. Sironi, C.F.: Monte-Carlo tree search for artificial general intelligence in games. Ph.D. thesis, Maastricht University (2019)

22. Soemers, D.J.N.J., Sironi, C.F., Schuster, T., Winands, M.H.M.: Enhancements for real-time Monte-Carlo tree search in general video game playing. In: 2016 IEEE Conference on Computational Intelligence and Games (CIG), pp. 436–443. IEEE (2016)

23. Tak, M.J.W., Winands, M.H.M., Björnsson, Y.: N-grams and the last-good-reply policy applied in general game playing. IEEE Trans. Comput. Intell. AI Games 4(2), 73–83 (2012)

24. Tak, M.J.W., Winands, M.H.M., Björnsson, Y.: Decaying simulation strategies. IEEE Trans. Comput. Intell. AI Games 6(4), 395–406 (2014)

Author Index

Cazenave, Tristan   1, 17, 56, 71, 84, 100, 116

Fournier, Thomas   84
Fürnkranz, Johannes   39

Jaouen, Adel   31
Joppen, Tobias   39

Le Merrer, Erwan   31

Negrevergne, Benjamin   100

Sevestre, Jean-Baptiste   17
Sikora, Florian   100
Sironi, Chiara F.   116

Toulemont, Matthieu   17

Ventos, Véronique   1

Winands, Mark H. M.   116

Printed in the United States
by Baker & Taylor Publisher Services